Strategic Planning Models for Reverse and Closed-Loop Supply Chains

Strategic Planning Models for Reverse and Closed-Loop Supply Chains

Kishore K. Pochampally
Satish Nukala
Surendra M. Gupta

CRC Press
Taylor & Francis Group
Boca Raton London New York

CRC Press is an imprint of the
Taylor & Francis Group, an **informa** business

CRC Press
Taylor & Francis Group
6000 Broken Sound Parkway NW, Suite 300
Boca Raton, FL 33487-2742

First issued in paperback 2019

© 2009 by Taylor & Francis Group, LLC
CRC Press is an imprint of Taylor & Francis Group, an Informa business

No claim to original U.S. Government works

ISBN-13: 978-1-4200-5478-1 (hbk)
ISBN-13: 978-0-367-38683-2 (pbk)

Library of Congress Cataloging-in-Publication Data

Pochampally, Kishore K.
    Strategic planning models for reverse and closed-loop supply chains / Kishore K. Pochampally, Satish Nukala, Surendra M. Gupta.
        p. cm.
    Includes bibliographical references and index.
    ISBN 978-1-4200-5478-1 (hardback : alk. paper)
    1. Manufacturing processes--Environmental aspects. 2. Business logistics--Environmental aspects. 3. Remanufacturing. 4. Waste minimization. 5. Salvage (Waste, etc.) 6. Recycling (Waste, etc.) I. Nukala, Satish. II. Gupta, Surendra M. III. Title.

TS155.7.P63 2008
658.5'67--dc22   2008014566

Visit the Taylor & Francis Web site at
http://www.taylorandfrancis.com

and the CRC Press Web site at
http://www.crcpress.com

Dedication

To our families:

Seema Dasgupta, Sharada Pochampally, and Narasimha Rao Pochampally

—KKP

Somayajulu Nukala, Jagadeeswari Nukala, Sridhar Nukala,

Mythili Nukala, and Padmaja Nukala

—SN

Sharda Gupta, Monica Gupta, and Neil Gupta

—SMG

Contents

Preface

The rapid technological development of new products, along with the growing desire of consumers to acquire the latest technology, has led to a new environmental problem: products that are discarded prematurely. But behind every problem lies an opportunity. In this case, the opportunity comes from *reprocessing* (processing used products), which leads to:

1. *Saving natural resources*: We conserve land and reduce the need to drill for oil and dig for minerals by making products using materials and components obtained from reprocessing rather than virgin resources.
2. *Saving energy*: It usually takes less energy to make products from reprocessed materials and components than from virgin resources.
3. *Saving clean air and water*: Making products from reprocessed materials and components creates less air and water pollution than from virgin resources.
4. *Saving landfill space*: When reprocessed materials and components are used to make a product, they do not go into landfills.
5. *Saving money*: It costs much less to make products from reprocessed materials and components than from virgin resources.

Besides the above opportunities, an important driver for companies to engage in reprocessing is the enforcement of environmental regulations by local governments.

A reverse supply chain consists of a series of activities required to collect used products from consumers and reprocess them to either recover their leftover market values or properly dispose of them. Today, in practice, it has become common for companies involved in a forward supply chain (series of activities required to produce new products from virgin resources and distribute them to consumers) to also carry out collection and reprocessing of used products. This combined practice of forward and reverse supply chains is called a closed-loop supply chain. In the past decade, there has been an explosive growth of reverse and closed-loop supply chains, in both scope and scale.

Strategic planning (also called designing) primarily involves the structuring (which products should be processed/produced in which facilities) of a supply chain over the next several years. It is long-range planning and is typically performed every few years when a supply chain needs to expand its capabilities. The issues faced by strategic planners of reverse and closed-loop supply chains are evaluation and selection of new and used products, collection centers, recovery facilities, marketing strategies, and production facilities, as well as evaluation of the futurity of used products, selection of secondhand markets, optimization of transportation of goods, synchronization of supply chain processes, and supply chain performance measurement. In all of these issues, strategic planners must meet the following challenges: uncertainty in supply rate of used products, unknown condition of used products, and imperfect correlation between supply of used products and demand for reprocessed goods.

This book addresses the above issues amidst the above challenges in a variety of decision-making situations using efficient models. These models implement several quantitative techniques, such as analytic hierarchy process, eigen vector method, analytic network process, fuzzy logic, extent analysis method, fuzzy multicriteria analysis method, quality function deployment, method of total preferences, linear physical programming, goal programming, technique for order preference by similarity to ideal solution (TOPSIS), Borda's choice rule, expert systems, Bayesian updating, Taguchi loss function, Six Sigma, neural networks, geographical information systems, and linear integer programming.

The issues addressed in this book can serve as foundations for other researchers to build bodies of knowledge in this new and fast-growing field of research, viz., strategic planning of reverse and closed-loop supply chains. Furthermore, the models proposed in this book for those issues can be utilized by industrialists for understanding how a particular issue in the strategic planning of reverse and closed-loop supply chains can be effectively approached in a particular decision-making situation, using a suitable quantitative technique or a suitable combination of two or more quantitative techniques.

The strategic planning of reverse and closed-loop supply chains remains an important and promising field of research. This is desirable from both environmental and economical points of view, because by using innovations in this field, environmentalists as well as industrialists can fully exploit their desired goals. The authors express their hope that this book will inspire further research and motivate new and rewarding research in this all too important field of study.

Acknowledgments

Encouragement for writing this book came from many sources. We thank hundreds of researchers from whose work we have benefited and many of whom we have had the good fortune of meeting and interacting with at conferences around the world. Many thanks to Dr. Sagar V. Kamarthi (Northeastern University), Dr. Elif Kongar (University of Bridgeport), Dr. Martin Bradley (Southern New Hampshire University), Dr. Tej S. Dhakar (Southern New Hampshire University), and Dr. Seamus M. McGovern (National Transportation Systems Center).

This book would not have been possible without the commitment of Taylor & Francis for encouraging innovative ideas. We express our appreciation to Taylor & Francis and its staff for providing seamless support in making it possible to complete this timely and important book.

Most importantly, we are indebted to our families, to whom this book is lovingly dedicated, for constantly providing us their unconditional support in making this book a reality.

Kishore K. Pochampally, PhD
Manchester, New Hampshire

Satish Nukala, PhD
Houston, Texas

Surendra M. Gupta, PhD
Boston, Massachusetts

About the Authors

Dr. Kishore K. Pochampally is an assistant professor of quantitative studies and operations management at Southern New Hampshire University in Manchester. His prior academic experience is as a postdoctoral fellow at Massachusetts Institute of Technology (MIT) in Cambridge. He holds a PhD in industrial engineering from Northeastern University in Boston. His research interests are in the areas of reverse logistics, supply chain design, and Six Sigma quality management. He is a Six Sigma Green Belt and a Project Management Professional (PMP®).

Dr. Satish Nukala is a senior supply chain analyst with Halliburton Energy Services in Houston, Texas. He holds a PhD in industrial engineering from Northeastern University in Boston and an MS in industrial and systems engineering from the University of Alabama in Huntsville. He is a Six Sigma Green Belt. His research interests are in the areas of production and operations management and total quality management.

Dr. Surendra M. Gupta, PE, is a professor of mechanical and industrial engineering and director of the Laboratory for Responsible Manufacturing at Northeastern University in Boston. He received his BE in electronics engineering from Birla Institute of Technology and Science, MBA from Bryant University, and MSIE and PhD in industrial engineering from Purdue University. Dr. Gupta's research interests are in the areas of production/manufacturing systems and operations research. He is mostly interested in environmentally conscious manufacturing, electronics manufacturing, MRP, JIT, and queueing theory. He has authored and coauthored more than 350 technical papers published in prestigious journals, books, and conference proceedings. His publications have been cited by thousands of researchers all over the world in journals, proceedings, books, and dissertations. He has traveled to all seven continents and presented his work at international conferences there (except Antarctica). He is currently serving as the area editor of "Environmental Issues" for *Computers and Industrial Engineering*, associate editor for *International Journal of Agile Systems and Management*, and editorial board member of a variety of journals. He has also served as conference chair, track chair, and member of technical committees of a variety of international conferences. Dr. Gupta has been elected to the memberships of several honor societies and is listed in various *Who's Who* publications. He is a registered professional engineer in the state of Massachusetts and a member of ASEE, DSI, IIE, INFORMS, and POMS. Dr. Gupta is a recipient of the Outstanding Research Award and the Outstanding Industrial Engineering Professor Award (in recognition of teaching excellence) from Northeastern University. His recent activities can be viewed at http://www1.coe.neu.edu/~smgupta/, and he can be reached by e-mail at gupta@neu.edu.

1

Introduction

1.1 Motivation

The level of consumption by a growing population in this world of finite resources and disposal capacities has been continuously increasing. This is fueled by the growing desire of consumers to acquire the latest technology, both at home and in the workplace, along with the rapid technological development of new products. As a result, the world now faces a serious environmental problem: waste, viz., used products that are discarded prematurely. For example, an estimated 60 million computers enter the market in the United States every year and more than 12 million computers are discarded every year, out of which fewer than 10% are reprocessed,* while the rest head to landfills [1]. According to the Environmental Protection Agency's *Municipal Solid Waste Fact Book*, 29 states in the United States have 10 or more years of landfill space left, 15 states between 5 and 10 years, and 6 states fewer than 5 years [2]. Increased consumption results in increased use of raw material and energy, thereby depleting the world's finite natural resources. The data [3, 4] presented in tables 1.1 and 1.2 give an idea about the materials and energy consumed in the manufacturing of microchips and LCD monitors, respectively. Apart from the excessive consumption of the world's finite natural resources and disposal capacities, the presence of toxic material, such as lead, polybrominated diphenyl ether, mercury, and hexavalent chromium, in the discarded electronic equipment poses a serious threat to the environment. This environmental degradation is not sustainable by Earth's ecosystem [6].

Reprocessing of used products helps in:

---

* Though direct reuse of the used products is infeasible in most cases, *remanufacturing* and *recycling* are the major reprocessing options applied in the industry. Remanufacturing is a process in which used products are restored to like-new conditions, and recycling is a process performed to retrieve the material content of used products without retaining the identity of their components [5].

TABLE 1.1

Materials and energy used in manufacturing microchips

Material	Description	Amount per memory chip	Annual use by industry worldwide	Amount used to make chips in one computer
Silicon wafer		0.25 g	4,400 tons	0.025 kg
Chemicals	Dopants	0.016 g	280 tons	0.002 kg
	Photolithography	22 g	390,000 tons	2.2 kg
	Etchants	0.37 g	6,600 tons	0.037 kg
	Acids/bases	50 g	890,000 tons	4.9 kg
	Total chemicals	72 g	1.3 million tons	7.1 kg
Elemental gases	N_2, O_2, H_2, He, Ar	700 g	12 million tons	69 kg
Energy	Electricity	2.9 kWh	52 billion kWh	281 kWh
	Direct fossil fuels	1.6 MJ	28 billion MJ	155 MJ
	Embodied fossil fuels	970 g	17 million tons	94 kg
Water		32 liters	570 billion liters	310 liters

TABLE 1.2

Aggregate chemicals, energy, and water use in manufacture of LCD monitor

Material/input	Amount used per monitor
Photolithographic and other chemicals	3.7 kg
Elemental gases (N_2, O_2, Ar)	5.9 kg
Electricity	87 kWh
Direct fossil fuels (98% natural gas)	198 kg
Embodied fossil fuels	226 kg
Water	1,290 liters

1. *Saving natural resources*: We conserve land and reduce the need to drill for oil and dig for minerals by making products using materials and components obtained from reprocessing instead of virgin materials.

2. *Saving energy*: It usually takes less energy to make products from reprocessed materials and components than from virgin materials.

3. *Saving clean air and water*: Making products from reprocessed materials and components creates less air pollution and water pollution than from virgin materials.

4. *Saving landfill space*: When reprocessed materials and components are used to make a product, they do not go into landfills.

5. *Saving money*: It costs much less to make products from reprocessed materials and components than from virgin materials.

As waste reduction is becoming a major concern in industrialized countries, the concept of reprocessing is gradually replacing a one-way perception of economy [7]. Increasingly, customers are expecting companies to minimize the environmental impact of their products and processes. Moreover, legislation extending producers' responsibility has become an important element of public environmental policy. Several countries, particularly in the European Union, have introduced environmental legislation charging manufacturers with responsibility for the whole life cycle of their products. The obligations of collection of used products and their reprocessing have been enacted or are under way for a number of product categories, including electronic equipment in the European Union and Japan, cars in the European Union and Taiwan, and packaging material in Germany. Also, companies are recognizing opportunities for combining environmental stewardship with profitability, brought about by production cost savings and access to new market segments. In this vein, in the past decade, there has been an explosive growth of reprocessing activities in both scope and scale. However, in companies in the United States, reprocessing is still in its infancy. In the United States, cities and towns are responsible for retrieval of used products and proper disposal of the potentially environmentally dangerous and waste components. According to a 2003 report [8], in the state of Massachusetts, support is building for a refiled bill that would require manufacturers of electronic products to pay for collection and reprocessing of those products after the end of their usage by the consumers. If passed, the statewide take-back program would be the first of its kind in the nation and would relieve cities and towns, which are bracing for local aid cuts, from the costs associated with collecting and disposing of electronic waste. The bill's supporters say that cities and towns in the United States spend between $6 million and $21 million a year on such endeavors.

A reverse supply chain (also known as *reverse logistics*) consists of a series of activities required to collect a used product from a consumer and reprocess it (used product) to either recover its leftover market value or dispose of it. Implementation of any reverse supply chain requires at least three parties: collection centers where consumers return used products, recovery facilities where reprocessing (remanufacturing or recycling) is performed, and demand centers where customers buy reprocessed products, viz., outgoing goods from recovery facilities. Figure 1.1 shows a generic reverse supply chain.

Environmental consciousness suggests the production of new products from conceptual design to final delivery such that the environmental standards and requirements are satisfied. In the last decade, environmental consciousness has become an obligation to many facilities in a traditional/forward supply chain (i.e., series of activities required to produce new products from raw material and distribute the former to customers), enforced primarily by governmental regulations and customer perspective on environmental issues [9, 10]. At the same time, many of these facilities are driven, mainly by profitability, to administer the reverse supply chain as well. The combination of forward and reverse supply chains is called a closed-loop supply chain. A generic closed-loop supply chain is shown in figure 1.2.

FIGURE 1.1
Generic reverse supply chain.

Reverse/closed-loop supply chains are implemented in many industries, including commercial aircraft, automobiles, computers, chemicals, appliances, and apparel [11].

A supply chain involves three phases of decision making: supply chain design, supply chain planning, and supply chain operation [12]:

- *Supply chain design*: This phase is also called *strategic planning*. During this phase, a company decides how to structure its supply chain over the next several years. Decisions made by companies include the location and capacity of production and warehouse facilities, the products to be manufactured and stored at various locations, the modes of transportation to be made available, and the type of information system to be utilized. Supply chain design decisions are typically made for the long term (years) and are very expensive to alter on short notice. Consequently, when companies make these decisions, they must take into account uncertainty in anticipated market conditions over the next few years.

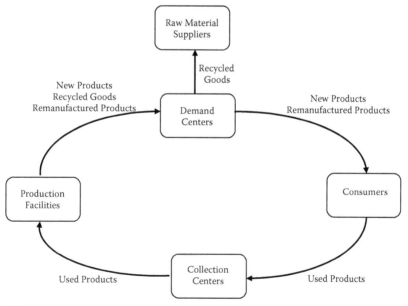

FIGURE 1.2
Generic closed-loop supply chain.

- *Supply chain planning*: This phase is also called *tactical planning*. Supply chain planning includes decisions regarding subcontracting of manufacturing, inventory policies to be followed, and timing and size of marketing promotions. For decisions made during this phase, the time frame considered is 3 months to a year. Companies must include uncertainty in demand, exchange rates, and competition over the time frame. Given a shorter time frame and better forecasts than the design phase, companies in this phase try to incorporate any flexibility built into the supply chain in the design phase and exploit it to optimize performance.

- *Supply chain operation*: This phase is also called *operational planning*. During this phase, companies make decisions regarding individual customer orders. The goal of supply chain operations is to handle incoming customer orders in the best possible manner. Decisions include allocating inventory or production to individual orders, setting a date for an order to be filled, generating pick lists at a warehouse, setting delivery schedules of trucks, and placing replenishment orders. Because operational decisions are made in the short term (minutes, hours, or days), there is less uncertainty about demand information.

This book focuses on strategic planning of reverse and closed-loop supply chains. The various decision-making problems faced by strategic planners of reverse and closed-loop supply chains include selection of used products to reprocess, evaluation of collection centers, evaluation of recovery facilities, selection of new products to produce, and optimization of transportation of goods.

Reverse and closed-loop supply chains differ from traditional/forward supply chains in many aspects and are complex to handle because of the inherent uncertainty involved in every stage of strategic planning (see table 1.3 [13]). As a result, strategic planning models for a forward supply chain cannot be adopted for a reverse or closed-loop supply chain.

This chapter is organized as follows: section 1.2 presents the overview of this book, section 1.3 gives the outline of the book, and section 1.4 gives some conclusions.

1.2 Overview of the Book

This book presents quantitative models for various issues faced by strategic planners of reverse and closed-loop supply chains amidst many challenges, such as uncertainty in supply rate of used products, unknown condition of used products, and imperfect correlation between supply of used products and demand for reprocessed goods.

TABLE 1.3

Comparison between forward and reverse/closed-loop supply chains

Forward supply chain	Reverse/closed-loop supply chain
Product quality uniform	Product quality not uniform
Disposition options clear	Disposition options unclear
Routing of products unambiguous	Routing of products ambiguous
Costs involved easily understood	Costs involved not easily understood
Product pricing uniform	Product pricing not uniform
Inventory management consistent	Inventory management inconsistent
Product life cycle manageable	Product life cycle less manageable
Financial management issues clear	Financial management issues unclear
Negotiations between parties straightforward	Negotiations less straightforward
Customer easily identifiable to market	Customer less easily identifiable to market
Forecasting relatively straight forward	Forecasting more difficult
One-to-many transportation	Many-to-one transportation
Marketing methods well known	Marketing complication by several factors
Process visibility more transparent	Process visibility less transparent

The complete list of issues addressed in this book is the following (see chapter 2 for further explanation of each of the issues):

- Selection of used products
- Evaluation of collection centers
- Evaluation of recovery facilities
- Optimization of transportation of goods
- Evaluation of marketing strategies
- Evaluation of production facilities
- Evaluation of futurity of used products
- Selection of new products
- Selection of secondhand markets
- Synchronization of supply chain processes
- Supply chain performance measurement

The models for addressing the above strategic planning issues employ several quantitative techniques, such as:

- Analytic hierarchy process
- Eigen vector method
- Analytic network process
- Fuzzy logic

- Extent analysis method
- Fuzzy multicriteria analysis method
- Quality function deployment
- Method of total preferences
- Linear physical programming
- Goal programming
- Technique for order preference by similarity to ideal solution (TOPSIS)
- Borda's choice rule
- Expert systems
- Bayesian updating
- Taguchi loss function
- Six Sigma
- Neural networks
- Geographical information systems
- Linear integer programming

The issues addressed in this book can serve as foundations for other researchers to build bodies of knowledge in this new and fast-growing field of research, i.e., strategic planning of reverse and closed-loop supply chains. Furthermore, the models proposed in this book for those issues can be utilized by industrialists for understanding how a particular issue in the strategic planning of reverse and closed-loop supply chains can be effectively approached in a particular decision-making situation, using a suitable quantitative technique or a suitable combination of two or more quantitative techniques.

1.3 Outline of the Book

This book is organized as follows:

In chapter 2, an overview of the strategic planning issues (see section 1.2) is presented, and it is explained how the decision-making situations could differ for each of the issues.

In chapter 3, a brief review of literature in the area of strategic planning of reverse and closed-loop supply chains is presented.

In chapter 4, each of the several quantitative techniques that are used in the strategic planning models presented in this book is briefly described.

In chapter 5, two models for selection of used products in two different decision-making situations are presented. The first model employs linear integer programming, and the second model employs linear physical programming.

In chapter 6, five models for evaluation of collection centers in five different decision-making situations are presented. The first model employs eigen vector method and Taguchi loss function. The second model uses eigen vector method, technique for order preference by similarity to ideal solution (TOPSIS), and Borda's choice rule. The third model uses neural networks, fuzzy logic, TOPSIS, and Borda's choice rule. The fourth model uses analytic network process (ANP) and goal programming. The fifth model uses eigen vector method, Taguchi loss function, and goal programming.

In chapter 7, five models for evaluation of recovery facilities in five different decision-making situations are presented. The first model uses analytic hierarchy process. The second model employs linear physical programming. The third model uses eigen vector method, technique for order preference by similarity to ideal solution (TOPSIS), and Borda's choice rule. The fourth model uses neural networks, fuzzy logic, TOPSIS, and Borda's choice rule. The fifth model uses a simple two-dimensional chart.

Chapter 8 focuses on achieving transportation of the right quantities of products (used, remanufactured, and new) across a reverse or closed-loop supply chain, while satisfying certain constraints. Five models are presented in five different decision-making situations. The first model employs linear integer programming. The second model employs linear physical programming. The third and fourth models use goal programming and linear physical programming, respectively. The fifth model employs fuzzy goal programming.

In chapter 9, three models for evaluation of the marketing strategy of a reverse/closed-loop supply chain in three different decision-making situations are presented. The first model employs fuzzy logic and technique for order preference by similarity to ideal solution (TOPSIS). The second model employs fuzzy logic, quality function deployment (QFD), and method of total preferences. The third model uses fuzzy logic, extent analysis method, and analytic network process.

In chapter 10, three models for evaluation of production facilities in five different decision-making situations are presented. The first model employs fuzzy logic and technique for order preference by similarity to ideal solution (TOPSIS). The second model employs fuzzy logic, extent analysis method, and analytic network process. The third model uses fuzzy multicriteria analysis method.

In chapter 11, it is shown how an expert system can be built using Bayesian updating and fuzzy logic, to decide whether it is more sensible to repair a used product of interest for subsequent sale on a secondhand market than to disassemble it for subsequent reprocessing.

In chapter 12, a fuzzy cost-benefit function is formulated and then used to perform a multicriteria economic analysis for selecting an economical new product to produce in a closed-loop supply chain.

In chapter 13, fuzzy logic, quality function deployment (QFD), and method of total preferences are used to select the market with the most potential in which to sell a used product from a set of candidate secondhand markets.

In chapter 14, a model consisting of two design experiments that use the Six Sigma concept to achieve better synchronization in a reverse supply chain is presented. This model tailors the individual processes in such a way that the overall delivery performance is maximized.

In chapter 15, appropriate performance aspects and their enablers (drivers of performance metrics) are identified for a reverse/closed-loop supply chain environment, and a performance measurement model that uses linear physical programming (LPP) and quality function deployment (QFD) is presented.

Chapter 16 presents the conclusions of this book.

1.4 Conclusions

This chapter first presented the factors that motivated this project, viz., drivers of reverse and closed-loop supply chains. Then the overview and outline of the book were presented.

References

1. Platt, B., and Hyde, J. 1997. *Plug into electronics reuse*, 13–38. Washington, DC: Institute for Local Self Reliance.
2. Knemeyer, A. M., Ponzurick, G. T., and Logar, M. C. 2002. A qualitative examination of factors affecting reverse logistics systems for end-of-life computers. *International Journal of Physical Distribution and Logistics Management* 32:455–79.
3. Socolof, M., Overly, L. K., and Greibig, J. 2003. *Desktop computer displays: A life-cycle assessment*. EPA 744-R-01-004. Washington, DC: U.S. Environmental Protection Agency.
4. Williams, E., Ayres, R., and Heller, M. 2002. The 1.7 kg microchip: Energy and chemical use in the production of semiconductors. *Environmental Science and Technology* 36:5504–10.
5. Gungor, A., and Gupta, S. M. 1999. Issues in environmentally conscious manufacturing and product recovery: A survey. *Computers and Industrial Engineering* 36:811–53.
6. Beamon, M. B., and Fernandes, C. 2004. Supply-chain network configuration for product recovery. *Production Planning and Control* 15:270–81.
7. Fleischmann, M. 2001. *Quantitative models for reverse logistics: Lecture notes in economics and mathematical systems*. Berlin: Springler-Verlag.
8. Anonymous. 2003. Computer recycling bill is gaining support. *Boston Metro*, January 10–12, p. 6.
9. Lambert, A. J. D., and Gupta, S. M. 2005. *Disassembly modeling for assembly, maintenance, reuse, and recycling*. Boca Raton, FL: CRC Press.

10. Savaskan, R. C., Bhattacharya, S., and Van Wassenhove Luk, N. 2004. Closed-loop supply chain models with product remanufacturing. *Management Science* 50:239–52.
11. Thierry, M., Salomon, M., van Nunen, J., and van Wassenhove, L. N. 1995. Strategic issues in product recovery management. *California Management Review* 37:114–35.
12. Chopra, S., and Meindl, P. 2006. *Supply chain management: Strategy, planning and operations*. Englewood Cliffs, NJ: Prentice Hall.
13. Rogers, D. 2001. RLEC project plans. *Livonia*, October, p. 18.

2

Strategic Planning of Reverse and Closed-Loop Supply Chains

2.1 Introduction

As explained in chapter 1, the first decision-making phase in a supply chain is *supply chain design*, which is also called *strategic planning*. Strategic planning primarily involves the structuring of a supply chain over the next several years. It is long-range planning and is typically performed every few years when a supply chain needs to expand its capabilities [1].

The various issues faced by strategic planners of reverse and closed-loop supply chains are the following:

1. Selection of used products
2. Evaluation of collection centers
3. Evaluation of recovery facilities
4. Optimization of transportation of goods
5. Evaluation of marketing strategies
6. Evaluation of production facilities
7. Evaluation of futurity of used products
8. Selection of new products
9. Selection of secondhand markets
10. Synchronization of supply chain processes
11. Supply chain performance measurement

In all of these issues, strategic planners must meet the following challenges: uncertainty in supply rate of used products, unknown condition of used products, and imperfect correlation between supply of used products and demand for reprocessed goods.

This chapter gives an overview of the aforementioned issues (sections 2.2 through 2.12, respectively) and how they could be addressed in a variety of decision-making situations using effective quantitative models. Section 2.13 gives some conclusions.

2.2   Selection of Used Products

Although many original equipment manufacturers (OEMs) are obligated to take products back from the consumers upon the products' end of use (hence called *used products*), there are also many third-party companies that collect used products solely to make profit. These companies select only those used products for which revenues from recycle or resale of the products' components are expected to be higher than the costs involved in collection and reprocessing of used products and in disposal of waste. The various scenarios for selecting economical used products could differ as follows:

1. Evaluation criteria could be presented in terms of classical numerical constraints.
2. Evaluation criteria could be presented in terms of ranges of different degrees of desirability.

2.3   Evaluation of Collection Centers

Designing an efficient reverse or closed-loop supply chain requires selection of efficient collection centers where used products are disposed of by the consumers. These collection centers, after initial processing (for example, sorting), ship the used products to recovery facilities or production facilities where reprocessing operations such as disassembly and recycling/remanufacturing are carried out.

The various scenarios for evaluating collection centers for efficiency could differ as follows:

1. Supply chain company executives, whose primary concern is profit, could be the sole decision makers.
2. There could exist three different categories of decision makers: consumers, local government officials, and supply chain company executives. The weights (importance values) of evaluation criteria are given.
3. There could exist the same three different categories of decision makers given in scenario 2, but weights of evaluation criteria are not given (hence, must be derived).
4. Evaluation could be made from the perspective of a remanufacturing facility interested in buying used products from the candidate collection centers. The goals are expressed in terms of performance indices (efficiency scores).

5. Evaluation could be made in the same manner as in scenario 4, but the goals are expressed in terms of Taguchi losses (inefficiency scores).

2.4 Evaluation of Recovery Facilities

Efficient recovery facilities (where reprocessing operations such as disassembly and recycling/remanufacturing are carried out) are as essential for a reverse supply chain as efficient collection centers. Hence, in addition to evaluation of collection centers (see section 2.3), strategic planning of a reverse supply chain involves evaluation of recovery facilities. The various scenarios for evaluating recovery facilities for efficiency could differ as follows:

1. Evaluation criteria could be given numerical weights (importance values).
2. Evaluation criteria could be presented in terms of ranges of different degrees of desirability.
3. Decision makers could have conflicting criteria for evaluation, and weights for evaluation criteria are given.
4. Decision makers could have conflicting criteria for evaluation, and weights for evaluation criteria are not given (hence, must be derived).
5. A very simple evaluation technique could be desired (where only the "most important" evaluation criteria are considered).

2.5 Optimization of Transportation of Goods

The focus of this issue is on achieving transportation of the right quantities of products (used, remanufactured, and new) across a reverse or closed-loop supply chain while satisfying certain constraints. The various scenarios for this problem could differ as follows:

1. Decision-making criteria for a reverse supply chain could be given in terms of classical supply-and-demand constraints.
2. Decision-making criteria for a reverse supply chain could be presented in terms of ranges of different degrees of desirability.
3. Besides optimal transportation of products, there could be a need to address the following issues in one continuous phase for a closed-loop supply chain: selection of used products and evaluation of

production facilities. Also, the decision-making criteria could be presented in terms of classic supply-and-demand constraints.

4. The scenario could be the same as scenario 3, except that the decision-making criteria could be presented in terms of different degrees of desirability.

5. The scenario could be the same as scenario 3, except that the decision-making criteria could be imprecise.

2.6   Evaluation of Marketing Strategies

A reverse/closed-loop supply chain program is successful only if there is a high level of public participation in the program. The level of public participation is shouldered by the marketing strategy of that program. Hence, evaluating the marketing strategy of a reverse/closed-loop supply chain program is equivalent to evaluating how well the strategy is driving the public to participate in the program. Whereas the drivers for governments and companies to implement a reverse/closed-loop supply chain program and evaluate the program's marketing strategies are environmental consciousness and profitability, respectively, the drivers for the public to participate in the program are numerous and often conflicting with each other (for example, the more regularly a reverse/closed-loop supply chain program offers to collect used products from consumers, the higher the taxes the consumers will have to pay). The various scenarios for evaluating the marketing strategy of a reverse/closed-loop supply chain program could differ as follows:

1. The program could be exclusively for reverse supply chain operations, i.e., absence of a closed loop.

2. The program could be for a closed-loop supply chain and the decision maker could be uninterested in considering interdependencies among evaluation criteria.

3. The program could be for a closed-loop supply chain and the decision maker could be interested in considering interdependencies among evaluation criteria.

2.7   Evaluation of Production Facilities

Efficient production facilities (not only new products are produced but also used products are reprocessed) are as essential for a closed-loop supply chain as efficient collection centers. Hence, in addition to selecting efficient collection centers (see section 2.3), strategic planning of a closed-loop supply chain involves evaluation of production facilities. The various scenarios for evaluating production facilities for efficiency could differ as follows:

1. The decision maker could desire to structure the problem using a simple hierarchical model, wherein interactions among evaluation criteria could be ignored.
2. Interactions among evaluation criteria could not be ignored, which in turn leads to a more complex problem structure.
3. Interactions among evaluation criteria could not be ignored, and the decision maker could desire to see how close the rating of a candidate production facility is to the "ideal solution."

2.8   Evaluation of Futurity of Used Products

Although a major driver for companies interested in collecting used products is recoverable value through reprocessing, those companies seldom know when those products were bought and why they were discarded. Also, the products do not indicate their remaining life periods. Hence, they often undergo partial or complete disassembly for subsequent reprocessing. The focus of this issue is to find out whether, for some products, it makes more sense to make necessary repairs to the products and sell them on second-hand markets than to disassemble them for subsequent reprocessing.

2.9   Selection of New Products

The focus of this issue is to help companies select and produce only those new products for which revenues in the closed-loop supply chain are expected to be higher than the costs. The revenues include new product sale revenue (revenue from selling new products, viz., products in the forward flow of the closed-loop supply chain), reuse revenue (revenue from direct sale/usage in remanufacturing usable components of used products), and recycle revenue (revenue from selling material obtained from recycling of unusable components of used products). The costs include new product production cost (cost to produce new products), collection cost (cost to collect used products from consumers), reprocessing cost (cost to remanufacture/recycle used products), disposal cost (cost to dispose of the material left over after remanufacturing or recycling of used products), loss-of-sale cost (cost due to loss of sale, which might occur occasionally, due to lack of supply of used products), and investment cost (capital required for facilities and machinery involved in production of new products and collection and reprocessing of used products).

2.10   Selection of Secondhand Markets

The focus of this issue is to select the market with the most potential in which to sell a *repaired* used product from a set of candidate secondhand markets. The main criteria for evaluation are before-sale performance (reflects the ability to attract new customers to the secondhand market), while-sale performance (reflects the ability to motivate the customers to buy secondhand products while the customers are *in* the secondhand market), and after-sale performance (reflects the ability to attract old customers to the secondhand market).

2.11   Synchronization of Supply Chain Processes

Synchronization among the internal business processes of a supply chain is essential for effective management of the supply chain. Synchronization in a supply chain means reducing the variability among the internal business processes or partners such that each stakeholder in the supply chain acts in a way that is appropriately timed with the actions of the other stakeholders [2]. The delivery performance of a supply chain is maximized largely by synchronizing the internal business processes such that the final product fits in the customer-specified delivery window with a very high probability.

The focus of this issue is to achieve synchronization of processes, such as procurement, inspection, disassembly, remanufacturing, transportation, and delivery, in a reverse supply chain.

2.12   Supply Chain Performance Measurement

The focus of this issue is to measure the performance of a reverse/closed-loop supply chain with respect to various metrics, such as on-time delivery, green image, service efficiency, and location of facilities. Due to the inherent differences in various aspects between forward and reverse/closed-loop supply chains (see chapter 1), the performance metrics and evaluation techniques used in a forward supply chain cannot be extended to a reverse/closed-loop supply chain.

2.13   Conclusions

This chapter gave an overview of various issues faced by strategic planners of reverse and closed-loop supply chains in various decision-making situations.

References

1. Chopra, S., and Meindl, P. 2006. *Supply chain management*. 3rd ed. New York: Prentice Hall.
2. Antony, J., Swarnkar, R., and Tiwari, M. K. 2006. Design of synchronized supply chains: A genetic algorithm based Six Sigma constrained approach. *International Journal of Logistics Systems and Management* 2:120–41.

3

Literature Review

3.1 Introduction

As explained in chapter 1, a supply chain involves three phases of decision making: *supply chain design* (also called *strategic planning*), *supply chain planning* (also called *tactical planning*), and *supply chain operation* (also called *operational planning*). In most of the literature for reverse and closed-loop supply chains, tactical planning is addressed in conjunction with strategic planning.

In this chapter, a brief review of the literature related to the three phases of decision making in reverse and closed-loop supply chains, viz., strategic planning, tactical planning, and operational planning, is presented.

This chapter is organized as follows: section 3.2 presents a review of literature about operational planning in reverse and closed-loop supply chains, section 3.3 gives a review of literature about strategic and tactical planning in reverse and closed-loop supply chains, and section 3.4 gives some conclusions.

3.2 Operational Planning of Reverse and Closed-Loop Supply Chains

Disassembly marks the first step in reprocessing of used products. Disassembly is defined as a systematic method of separating a product into its constituent parts, components, subassemblies or other groupings through a series of operations. Disassembly can be of two types: *nondestructive* or *destructive*. Nondestructive disassembly is the process of systematic removal of the constituent parts from an assembly in a manner by which there is no impairment of the parts during the process. On the contrary, destructive disassembly involves separating materials from an assembly with an objective of sorting the different material types for recycling. Disassembly can be *complete* or *partial*. In a complete disassembly, the used product is fully disassembled, whereas in partial disassembly, the used product is not fully disassembled, and only certain parts or assemblies are recovered. See [15],

[29], [48], and [59] for a comprehensive list of issues related to disassembly planning and scheduling.

Pnueli and Zussman [64] propose a methodology that includes identifying reprocessing options of all the parts of a product and improving the product's design.

Penev and De Ron [63] discuss the determination of optimal disassembly level and sequences of a used product, which provide conditions for the generation of profit while considering the environmental impact. In order to obtain the optimal disassembly process strategy, the authors utilize graph theory and cost analysis.

Lambert [50] proposes a methodology to identify the disassembly level in an economically optimal way. After disassembly, reusable parts/subassemblies are cleaned, refurbished, tested, and directed to the part/subassembly inventory for reprocessing operations.

Lambert and Gupta's book [49] on disassembly modeling presents disassembly in the context of the entire product life cycle. It examines disassembly on the intermediate level, incorporating design for disassembly, concurrent design, and reverse supply chain. The book also presents a comprehensive discussion of the theories and methodologies associated with disassembly, and the authors incorporate real-world case examples to explore the three main areas of application: assembly optimization, maintenance and repair, and reprocessing.

Gungor and Gupta [28] provide an exhaustive review of literature in the areas of environmentally conscious manufacturing and product recovery that includes several aspects of environmentally conscious manufacturing (design for environment, design for disassembly, etc.), common issues in product recovery (recycling, remanufacturing), disassembly process planning (production planning and inventory control issues), etc.

A disassembly line is perhaps the most suitable setting for disassembling large products. Gungor and Gupta [27] discuss the importance of a disassembly line and identify the critical issues and complexities and their effects on the disassembly line.

Boon et al. [14] study the economic viability of recycling infrastructure in the United States in relation to the electronics industry. They use a goal programming technique to study the sensitivity of several parties (disassemblers, recyclers, etc.) to different factors involved in the end-of-life processing.

Kongar and Gupta [40] and Imtanavanich and Gupta [34, 35] propose several multicriteria decision-making methodologies in a disassembly-to-order setting under different decision-making environments, which, when solved, provide the number of used products to be taken back for reuse, recycling, storage, or disposal to meet the demand for those products/components.

Guide et al. [22] identify the problems associated with production planning issues in a closed-loop supply chain and advocate that each closed-loop supply chain differs from every other closed-loop supply chain and a "one size fits all" approach does not work. Each type of system offers different managerial concerns. They apply Hayes and Wheelwright's product-process matrix framework to examine three cases representing remanufacture-to-stock, reassemble-to-order, and remanufacture-to-order environments, thereby extending the product-process matrix to include the insights on production planning in closed-loop supply chains.

Rubio et al. [70] review an extensive list of articles on reverse logistics published in the production and operations management field. They provide a database of articles published in the area of reverse logistics in the period 1995–2005 by exploring the topic, methodology, and techniques of analysis, as well as other relevant aspects of research.

Meade et al. [58] provide an extensive review of literature in the area of reverse logistics as well as an overview of definitions and research opportunities in this area.

Inventory control and production planning methods are thoroughly understood and well established for traditional forward supply chains. However, available techniques for a traditional forward supply chain are not transferable and adequate for a reverse supply chain. Guide and Srivastava [24] list the factors that complicate the inventory control and production planning in a reverse supply chain. These factors are as follows:

- Probabilistic recovery rate of parts from the used products, which implies a high degree of uncertainty in material planning
- Unknown conditions of recovered parts until inspected, thus leading to stochastic routings and lead times
- Part-matching problem during the assembly process
- Added complexity of a remanufacturing shop structure
- Uncertainty in supply rate of used products
- Problem of imperfect correlation between supply of used products and demand for reprocessed goods

Inventory control models in a reverse supply chain are required to keep track of various goods through the chains, viz., used products, remanufactured products, recycled goods, etc. Inventory models for repairable items and maintenance systems where failed machines are replaced by warm or cold spares carry some similarities to the reverse supply chain models [11, 16]. Some authors consider deterministic models in which supply-and-demand rates are constant. Mabini et al. [57] consider such a model with fixed setup costs for orders and remanufacturing and linear holding costs for used products and finished goods. Also, the model includes understock service-level constraint and machine sharing during the remanufacturing stage. Richter [68, 69] proposes a similar model under a different control policy where he provides formulation to calculate optimal values for control parameters and analyzes their dependence on the supply rate of used products. In a later paper, Richter [67] considers a mixture of two pure policies, total disposal and total remanufacturing in a deterministic system. In a more recent paper, Teunter [78] studies a deterministic system with continuous review and no lead times for outside procurement or remanufacturing. Although it is assumed that remanufactured products are as good as new ones, the associated inventory holding costs for remanufactured products are lower than those for new items. The author proposes an Economic Order Quantity (EOQ) control mechanism with fixed batch sizes and shows that in a given time period, the optimal batch size should be equal to 1 for either manufacturing or remanufacturing.

In repairable item inventory and remanufacturing systems, stochastic supply and demand occurrences are controlled by periodic and continuous review models. Periodic review models aim to minimize expected costs over a finite period and obtain optimal control policies under various assumptions. On the other hand, continuous review models monitor the inventory level incessantly and aim to find static control policies to obtain optimal control parameters to minimize the expected system costs. Simpson [75] considers a periodic review model for separate inventories for serviceable and recoverable parts. He shows that a three-parameter control policy that controls demand, remanufacturing, and disposal activities is optimal. Cohen et al. [17] consider a periodic review model in which used products can be placed in serviceable inventory directly. The authors assume that a fixed percentage of supplied products is returned to the manufacturer after a fixed lead time. The model optimizes the trade-off between holding and shortage costs. For a similar model, Kelle and Silver [38] utilize an integer program based on the net demand per period under chance constraints with fixed setup costs. Simpson's model [75] is extended by Inderfurth [36] by considering the effects of nonzero lead times for the reprocessing process and orders. He considers a push system for remanufacturing to avoid the storage of used products and shows that the difference between these two lead times is an important complexity factor of the system. Kiesmuller and Van der Laan [39] consider a periodically reviewed system in a finite horizon with supply rates dependent on demand rates. The authors assume deterministic lead times and an independent stochastic demand process. They show that dependent supply rates provide a better inventory control performance. Teunter and Vlachos [77] consider the impact of disposal on the discounted average cost. Their model considers deterministic and equal lead times for remanufactured and new items procurement.

Heyman [31] analyzes the continuous review inventory control case where incoming used products are disposed of whenever the inventory position reaches a predetermined level. The author assumes zero repair times and does not consider procurement lead times. In a following paper, Muckstadt and Isaac [60] present a production planning and inventory control model for remanufacturing. The authors develop an approximate control strategy with respect to reorder points and order quantities for a single-item product case where used products are remanufactured. They consider fixed lead times and no disposal of used products. Van der Laan et al. [83] consider a single-product, single-echelon production and inventory system with used products, remanufacturing, and disposal. In their paper, the authors compare the performance of three different procurement and inventory control strategies. Salomon et al. [72] and Van der Laan and Salomon [82] develop and compare PUSH and PULL strategies for a joint production and inventory system using both remanufactured and new parts. Korugan and Gupta [41] consider a two-echelon inventory system with stochastic lead times and arrival rates. They model the system as a PUSH-type make-to-stock open queueing network, where along with the manufacturing and remanufacturing processes, the demand arrivals are modeled as service. The authors report that increasing supply rates decrease the expected total cost, as the holding cost of a used product at the lower echelon is lower than the lost sales cost. Then Korugan and Gupta [42] look at the CONWIP problem of the hybrid production systems with mutually exclusive manufacturing and remanufacturing processes. The authors generalize dynamic and static routing methods by applying an adaptive kanban control policy. The experiments conducted by the authors demonstrate that dedicated kanban control performs better with reasonable utilization rates, whereas adaptive kanban policy always outperforms nonadaptive policies. Udomsawat and Gupta [80] focus on the production control aspect of the disassembly environment. The author adapts the kanban control mechanism to a disassembly line and develops a system that uses several types of kanbans attached to various components and subassemblies.

Applicability of traditional production planning and scheduling methods for remanufacturing systems is extremely limited. Therefore, either new methodologies have to be developed or classical methods have to be modified to manage the complications of the product reprocessing process [81]. A number of authors propose that material requirements planning (MRP) techniques utilize a reverse bill of materials (BOM) to allow the management and control of inventories in the remanufacturing environment [30, 47, 62]. Guide and Srivastava [23] propose a specific structure for MRP mechanics and evaluate a method to calculate safety stock for material recovery uncertainty. A multiobjective mathematical model formulation for a recycle-oriented manufacturing system is presented by Hoshino et al. [32]. The optimization model for the recycle-oriented manufacturing system is designed for a single product with m number of parts. Each part has three attributes associated with it: (1) reusable part, (2) not reusable but reproducible part, and (3) not reproducible or reusable part. The authors apply a goal programming approach with two objectives: profit maximization and recycling rate maximization. Guide [25] proposes scheduling using the drum–buffer–rope concept as an alternative to the MRP method. Guide et al. [26] evaluate various part order-release strategies in a remanufacturing environment using a simulation model. The authors focus on the problem of uncontrolled release of parts from disassembly operations, which may cause long queues at machine centers. This situation may increase lead times and their variability, making customer service levels decline. Various types of order-release strategies are discussed, viz., a level strategy, a batch strategy, a local order strategy, and a global order-release strategy, and the findings are mixed as to what is the best disassembly release mechanism. Ferrer and Whybark [19] describe the first fully integrated material planning system to facilitate managing a remanufacturing facility. The approach extends the methods used for material planning in a remanufacturing environment in several ways. First, it explicitly links the volume of used products with the volume of sales. Second, it uses the bill of material for each component directly, with no need for modification. Third, the system derives the need for parts and uses optimizing procedures to determine the disassembly schedule. Finally, part commonality and different yield factors are explicitly included.

Aksoy and Gupta [2] model a hybrid manufacturing system with distinct cells for disassembly, testing, and remanufacturing activities. The hybrid system is modeled as an open queueing network (OQN) with unreliable servers and finite buffers. Then Aksoy and Gupta [1] look at the effect of reusable rate variation on the performance of a remanufacturing system. The authors report that when the supply rate of used products is low, the total cost is relatively insensitive to the reusability rate. However, as the supply rate becomes higher, the reusability rate has a significant effect on the total cost. The higher the reusability rate, the lower the total cost. Also, mean process time (average remanufacturing time) is sensitive to the buffer size, breakdown rate, repair rate, and service rate of the stations in the remanufacturing system. Variability in the buffer sizes in the network causes a significant variation in the mean process time. The authors also examine the trade-off between expanding the buffer sizes and increasing the service capacity of machines [3, 4].

3.3   Strategic and Tactical Planning of Reverse and Closed-Loop Supply Chains

Strategic and tactical planning of reverse and closed-loop supply chains is a relatively new area of research, and hence only a few quantitative models and case studies have been reported in the literature. In this section, we present a brief survey of those models and studies (see Fleischmann's book [20] for a detailed survey of many of the case studies):

- Veerakamolmal and Gupta [84] propose a cost-benefit function and a measure called "design for disassembly index" that analyzes the trade-off between the costs and benefits of end-of-life disassembly to find the optimum cost-benefit ratio for end-of-life retrieval. Their study helps in comparing two product designs by assessing feasible combinations of components to be retrieved from a used product. Further, it compares the combination with the highest cost-benefit from one design with those from the others.

- Louwers et al. [55] consider the design of a reverse supply chain for carpet waste in Europe. A continuous location model in which all costs are considered volume dependent is proposed. The nonlinear model, when solved, determines the appropriate locations and capacities for the regional recovery facilities, taking into consideration transportation, investment, and processing costs.

- Ravi et al. [65] propose a balanced scorecard and analytic network process (ANP)–based approach to evaluate the alternative reverse logistics operations for used computers. It is a holistic approach that links financial and nonfinancial, tangible and intangible, and internal and external factors, for the selection of an alternative.

- Beamon and Fernandes [10] propose a multiperiod mixed-integer linear programming model to study a closed-loop supply chain where manufacturers produce new products and remanufacture used products. Their model addresses issues that include which warehouses and collection centers should be open and which warehouses should have sorting capabilities.

- Barros et al. [8] address the design of a supply chain for recycling sand from processing construction waste in the Netherlands. A four-level sand recycling network is considered: (1) crushing companies yielding sieved sand from construction waste, (2) regional depots specifying the pollution level and storing cleaned and half-cleaned sand, (3) treatment facilities cleaning and storing polluted sand, and (4) infrastructure projects where sand can be reused. The locations of the sand sources are known and their supply volumes are estimated based on historical data. The optimal number, capacities, and locations of depots and cleaning facilities are to be determined. The authors propose a multilevel capacitated facility location model for this problem formulated as a mixed-integer linear programming model and solved via iterative rounding of LP relaxations strengthened by valid inequalities. Listes and Dekker [53] propose a stochastic programming–based approach for the sand recycling network to account for the uncertainties. The stochastic model seeks to find a solution that is approximately balanced between some alternative scenarios identified by field experts.

- Savaskan et al. [74] study the problem of choosing the appropriate reverse channel structure for collecting the used products from customers. They compare three decentralized closed-loop supply chain models with the manufacturer collecting the used products, the retailer collecting the used products, and a third party collecting the used products. The models are compared with respect to the wholesale price, product return rate, and total supply chain profits.

- Ammons et al. [7] address carpet recycling in the United States. A logistics network that includes collection of used carpets from carpet dealerships, as well as separation of nylon and other reusable materials while landfilling the remainder, is investigated. Although the delivery sites for recovered materials are assumed to be known, the optimal number and location of both collection sites and processing plants for alternative configurations are to be determined. In addition, the amount of carpet collected from each site is to be determined. Facility capacity constraints are the main restrictions in view of the vast volume landfilled. The authors propose a multilevel capacitated mixed-integer linear programming model to address this problem.

- Biehl et al. [13] simulate a reverse logistics network for carpet recycling to manage highly variable return flows. They use an experimental design technique to study the effect of system design factors as well as environmental factors that affect the operational performance of such a reverse logistics network. From their study, the authors conclude that even with the design of an efficient reverse logistics network and use of sophisticated recycling technologies, return flows cannot meet demand for nearly a decade. They also discuss possible managerial options to address this problem, which include legal responses to require return flows and utilization of market incentives for carpet recycling.

- Hu et al. [33] present a cost-minimization model for a multi-time-step, multitype hazardous wastes reverse logistics system. The authors formulate a discrete-time analytical model that minimizes total hazardous waste reverse logistics costs subject to constraints, including business operating strategies and government regulations. The critical activities that include waste collection, storage, processing, and distribution are considered in their model. By using the proposed methodology, coupled with operational strategies, it is found that the total reverse logistics costs can be reduced by up to 49%.

- Lieckens and Vandaele [51] combine queuing models with the traditional reverse logistics location model and formulate mixed-integer linear programming models to determine which facilities to open while minimizing the total cost of investment, transportation, disposal, procurement, etc. By combining the queuing models, some dynamic aspects such as lead time and inventory positions, and the high degree of uncertainty associated with reverse logistics networks, are accounted for. With these extensions, the problem is defined as a mixed-integer nonlinear programming model. The model is presented for a single-product, single-level network, and several case examples are solved using genetic algorithms based on the technique of differential evolution.

- Alshamrani et al. [5] study the reverse logistics network for blood distribution with the American Red Cross, where containers in which blood is delivered from a central processing unit to customers in one time period are available for return to the central processing unit the following period. Containers not picked in the period following their delivery incur a penalty cost. This leads to a dynamic logistics planning problem, where in each period the vehicle dispatcher needs to design a multistop vehicle route while determining the number of containers to be picked up at each stop. A heuristic procedure is developed to solve the route design–pickup strategy problem.

- Vlachos et al. [85] address the capacity planning issues in remanufacturing facilities in reverse supply chains through a simulation model based on the principles of system dynamics methodology. Apart from considering economic issues, their study takes into account environmental issues such as take-back obligations and the "green image" effect on customer demand. The simulation model serves as an experimental tool that helps in evaluating long-term capacity planning policies using total supply chain profit as a measure of effectiveness.

- Lu and Bostel [56] study the facility location problem in a remanufacturing network that has a strong interaction between the forward and reverse flows of products. Remanufactured products are introduced as new ones into the forward flow. The authors assume that the demand is deterministic and the facilities are of three different types: producers, remanufacturing centers, and intermediate centers. The problem is modeled as a 0-1 mixed-integer programming problem that is solved using an algorithm based on Lagrangian heuristics.

- Spengler et al. [76] study the recycling networks for industrial by-products in the German steel industry. Recycling facilities with variable capacity levels and corresponding fixed and variable processing costs can be installed at a set of potential locations. Thus, one needs to determine which recycling processes or process chains to install at which locations and their capacity levels. The authors propose a multilevel warehouse location model with piecewise linear costs, which is used for optimizing several scenarios.

- Thierry et al. [79] propose a conceptual model for a closed-loop supply chain that addresses the situation of a manufacturing company collecting used products for recovery in addition to producing and distributing new products. The recovered products are sold under the same conditions as new ones to satisfy a given market demand. The distribution network encompasses three levels: plants, warehouses, and markets. All facilities are fixed externally, and hence no fixed costs are considered in the model. The objective is to determine the cost-optimal flow of goods in the network under the given capacity constraints. The problem is formulated as a linear programming model, which can be solved for optimality.

- Berger and Debaillie [12] address a situation similar to that of Thierry et al. [79] and propose a conceptual model for extending an existing forward supply chain with disassembly centers to allow for recovery of used products. The model is illustrated in a fictitious case of a computer manufacturer. Although the facilities in the forward supply chain are fixed, the number, locations, and capacities of the disassembly centers are to be determined. In a variant of this model, the recovery network is extended to another level by separating inspection and disassembly/repair centers. After inspection, rejected items are disposed of, whereas recoverable items are sent to the disassembly centers. To this end, the authors propose a multilevel capacitated mixed-integer linear programming model.

- Lim et al. [52] propose a mixed-integer programming model that takes into account multiperiod planning horizons with uncertainties for a product with modular design and multiproduct configurations. Besides maximizing overall profit, minimizing environmental impacts by minimizing energy consumption is considered.

- Reimer et al. [66] model the economics of electronics recycling from the perspective of recyclers, generators, and material processors, individually. They propose a nonlinear mixed-integer programming model for optimizing processing decisions in electronics recycling operations.

- Jayaraman et al. [37] analyze the logistics network of an electronic equipment remanufacturing company in the United States. The company's activities include collection of used products from customers, remanufacturing, and distribution of remanufactured products. In this network, the optimal number and locations of remanufacturing facilities, and the number of used products collected, are to be determined while considering investment, transportation, processing, and storage costs. The authors propose a multiproduct capacitated warehouse location mixed-integer linear programming model that is solved to optimality for different supply-and-demand scenarios.

- Krikke et al. [43] report a case study concerning the implementation of a remanufacturing process at a copier manufacturing company in the Netherlands. The reverse supply chain is subdivided into three main stages: (1) disassembly of return products to a fixed level; (2) preparation, which encompasses the inspection and replacement of critical components; and (3) reassembly of the remaining carcass together with repaired and new components into a remanufactured machine. Although the supplying processes and disassembly are fixed, optimal locations and flow of goods are to be determined for both the preparation and the reassembly operations. Based on a mixed-integer linear programming model, the optimal solution minimizing operational costs is compared with a number of preselected managerial solutions.

- Kroon and Vrijens [46] consider the design of a logistics system for reusable transportation packaging. More specifically, a closed-loop, deposit-based system is considered for collapsible plastic containers that can be rented as secondary packaging material. The actors involved in the system include a central agency owning a pool of reusable containers; a logistics service provider responsible for storing, delivering, and collecting the empty containers; senders and recipients of full containers; and carriers transporting full containers from sender to recipient. In addition to determining the number of containers required for running the system and an appropriate fee per shipment, a major question is where to locate the depots for empty containers. An additional requirement is balancing the number of containers at the depots. The problem is formulated as a mixed-integer linear programming model that is closely related to a classical uncapacitated warehouse location model.

- Salema et al. [71] point out that the majority of the quantitative models that exist in the area of reverse supply chain design are case specific and hence lack generality. To this end, they propose a generic reverse supply chain model that incorporates multiproduct management, capacity limits, and uncertainty in product demands and returns, and they propose a mixed-integer formulation that is solved using standard Branch and Bound (B&B) techniques.

- Krikke et al. [45] focus on commercial returns that have nothing to do with environmental legislation. These returns are increasing due to trends such as product leasing, catalog/Internet sales, shorter product replacement cycles, and increased warranty claims. The authors propose several options for a closed-loop supply chain: reusing the product as a whole, reusing the components, or reusing the materials.

- Listes [54] presents a generic stochastic model for the design of networks constituting both supply and return channels, organized in a closed-loop system. The model accounts for a number of alternative scenarios, which may be constructed based on critical levels of design parameters such as demand or returns. A decomposition approach is suggested to solve this model, based on the branch-and-cut procedure known as the integer L-shaped method.

- Bautista and Pereira [9] address reverse logistics problems arising in municipal waste management. The usual collection system in the European Union countries is composed of two phases. First, citizens leave their refuse at special collection areas where different types of waste (glass, paper, plastic, organic material) are stored in special refuse bins. Subsequently, each type of waste is collected separately and moved to its final destination (a recycling plant or refuse dump). The study focuses on the problem of locating these collection areas. The authors propose a genetic algorithm and a GRASP heuristic to solve the problem.

- Dowlatshahi [18] identifies the present state of theory in reverse logistics by formulating the propositions for strategic factors. The approach used is grounded theory development. The strategic factors are delineated and evaluated in terms of specific subfactors associated with each factor by the use of interview protocol and within the context of an in-depth analysis of two companies in different industries that are engaged in remanufacturing/recycling operations within reverse logistics systems. Based on these insights and strategic factors and subfactors, a framework for effective design and implementation of remanufacturing/recycling operations in reverse logistics is provided. This framework allows for the determination of the viability of used products/parts.

- Amini et al. [6] discuss the competitive value of service management activities, particularly repair services, as well as the importance of the supporting role of effective reverse logistics operations for the successful and profitable execution of repair service activities. Also, they present a case study of a major international medical diagnostics manufacturer to illustrate how a reverse logistics operation for a repair service supply chain was designed for both effectiveness and profitability by achieving a rapid-cycle time goal for repair service while minimizing total capital and operational costs.

- Savaskan and Van Wassenhove [73] focus on the interaction between a manufacturer's *reverse channel* choice to collect postconsumer goods and the strategic product pricing decisions in the forward *channel* when retailing is competitive. To this end, they model a direct product collection system, in which the manufacturer collects used products directly from the consumers, and an indirect product collection system, in which the *retailers* act as product return points. The authors show that the buy-back payments transferred to the *retailers* for postconsumer goods provide a wholesale pricing flexibility that can be used to price-discriminate between *retailers* of different profitability.

- Wojanowski et al. [86] study the interplay between industrial firms and government concerning the *collection* of used products from households. The authors focus on the use of a *deposit-refund* requirement by the government when the *collection* rate voluntarily achieved by the firms is deemed insufficient. A continuous modeling framework is presented for designing a drop-off facility *network* and determining the sales price that maximizes the firm's profit *under* a given *deposit-refund*. The customers' preferences with regards to purchasing and returning the product are incorporated via a discrete choice model with stochastic utilities.

- Krikke et al. [44] develop quantitative modeling to support decision making concerning both the design structure of a product and the design structure of the logistic network. Environmental impacts are measured by linear energy and waste functions. Economic costs are modeled as linear functions of volumes with a fixed setup component for facilities. This model is applied to a closed-loop supply chain design problem for refrigerators using real-life data of a Japanese consumer electronics company.

- Nagurney and Toyasaki [61] develop an integrated framework for the modeling of reverse supply chain management of electronic waste, which includes recycling. The decision makers consist of the sources of electronic wastes, the recyclers, the processors, and consumers associated with the demand markets for the distinct products. A multitiered electronic recycling network equilibrium model is constructed. The authors also establish the variational inequality formulation. The variational inequality formulation allows for the formulation of the complex reverse supply chain network to obtain the endogenous equilibrium prices and material flows between tiers.

- Gautam and Kumar [21] describe a multiobjective evaluation of the trade-offs among the number and size of drop-off stations, population covered in a service network, average walking distance to drop-off stations by the population, and the distance traveled by collection vehicles. A geographical information system (GIS)–based model is proposed for the design of a solid-waste system considering waste generation, allocation, recycling options, and location of drop-off stations.

3.4 Conclusions

In this chapter, a brief review of the literature related to the three phases of decision making in reverse and closed-loop supply chains—strategic planning, tactical planning, and operational planning—was presented.

References

1. Aksoy, H. K., and Gupta, S. M. 2000. Effect of reusable rate variation on the performance of remanufacturing systems. In *Proceedings of the SPIE International Conference on Environmentally Conscious Manufacturing*, Boston, pp. 13–20.
2. Aksoy, H. K., and Gupta, S. M. 1999. An open queueing network model for remanufacturing systems. In *Proceedings of the 25th Conference on Computers and Industrial Engineering*, New Orleans, pp. 62–65.
3. Aksoy, H. K., and Gupta, S. M. 2005. Buffer allocation plan for a remanufacturing cell. *Computers & Industrial Engineering*, vol. 48, No. 3, 657–677.

4. Aksoy, H. K., and Gupta, S. M. 2001. Capacity and buffer trade-offs in a remanufacturing system. In *Proceedings of the SPIE International Conference on Environmentally Conscious Manufacturing II*, Newton, MA, pp. 167–74.
5. Alshamrani, A., Mathur, K., and Ballou, R. H. 2007. Reverse logistics: Simultaneous design of delivery routes and return strategies. *Computers and Operations Research* 34:595–619.
6. Amini, M. M., Retzlaff-Roberts, D., and Bienstock, C. C. 2005. Designing a reverse logistics operation for short cycle time repair services. *International Journal of Production Economics* 96:367–80.
7. Ammons, J. C., Realff, M. J., and Newton, D. J. 1999. Carpet recycling: Determining the reverse production system design. *Polymer-Plastics Technology and Engineering* 38:547–67.
8. Barros, A. I., Dekker, R., and Scholten, V. 1998. A two-level network for recycling sand: A case study. *European Journal of Operational Research* 110:199–214.
9. Bautista, J., and Pereira, J. 2006. Modeling the problem of locating collection areas for urban waste management. An application to the metropolitan area of Barcelona. *Omega* 34:617–29.
10. Beamon, M. B., and Fernandes, C. 2004. Supply-chain network configuration for product recovery. *Production Planning and Control* 15:270–81.
11. Berg, M. A. 1980. A marginal cost analysis for preventive replacement policies. *European Journal of Operations Research* 4:136–42.
12. Berger, T., and Debaillie, B. 1997. Location of disassembly centers for re-use to extend an existing distribution network (in Dutch). Master's thesis, University of Leuven, Belgium.
13. Biehl, M., Prater, E., and Realff, M. J. 2007. Assessing performance and uncertainty in developing carpet reverse logistics systems. *Computers and Industrial Engineering* 34:443–63.
14. Boon, J. E., Isaacs, J. A., and Gupta, S. M. 2002. Economic sensitivity for end of life planning and processing of PCs. *Journal of Electronics Manufacturing* 11:81–93.
15. Brennan, L., Gupta, S. M., and Taleb, K. N. 1994. Operations planning issues in an assembly/disassembly environment. *International Journal of Operations and Production Planning* 14:57–67.
16. Cho, D. I., and Parlar, M. 1991. A survey of maintenance models for multi-unit systems. *European Journal of Operations Research* 51:1–23.
17. Cohen, M. A., Nahmias, S., and Pierskalla, W. P. 1980. A dynamic inventory system with recycling. *Naval Research Logistics Quarterly* 27:289–96.
18. Dowlatshahi, S. 2005. A strategic framework for the design and implementation of remanufacturing operations in reverse logistics. *International Journal of Production Research* 43:3455–80.
19. Ferrer, G., and Whybark, D. C. 2001. Material planning for a remanufacturing facility. *Production and Operations Management* 10:112–24.
20. Fleischmann, M. 2001. *Quantitative models for reverse logistics: Lecture notes in economics and mathematical systems*. Berlin: Springler-Verlag.
21. Gautam, A. K., and Kumar, S. 2005. Strategic planning of recycling options by multi-objective programming in a GIS environment. *Clean Technologies and Environmental Policy* 7:306–16.
22. Guide, Jr., V. D. R., Jayaraman, V., and Linton, J. D. 2003. Building contingency planning for closed-loop supply chains with product recovery. *Journal of Operations Management* 21:259–79.

23. Guide, Jr., V. D. R., and Srivastava, R. 1997. An evaluation of order release strategies in a remanufacturing environment. *Computers and Operations Research* 24:37–47.
24. Guide, Jr., V. D. R., and Srivastava, R. 1997. Repairable inventory theory: Models and applications. *European Journal of Operations Research* 102:1–20.
25. Guide, Jr., V. D. R. 1996. Scheduling using drum-buffer-rope in a remanufacturing environment. *International Journal of Production Research* 34:1081–91.
26. Guide, Jr., V. D. R., Srivastava, R., and Spencer, M. S. 1997. An evaluation of capacity planning techniques in a remanufacturing environment. *International Journal of Production Research* 35:67–82.
27. Gungor, A., and Gupta, S. M. 2002. Disassembly line in product recovery. *International Journal of Production Research* 40:2569–89.
28. Gungor, A., and Gupta, S. M. 1999. Issues in environmentally conscious manufacturing and product recovery: A survey. *Computers and Industrial Engineering* 36:811–53.
29. Gupta, S. M., and McLean, C. R. 1996. Disassembly of products. *Computers and Industrial Engineering* 31:225–28.
30. Gupta, S. M., and Taleb, K. N. 1994. Scheduling disassembly. *International Journal of Production Research* 32:1857–66.
31. Heyman, D. P. 1997. Optimal disposal policies for a single item inventory system with returns. *Naval Research Logistics Quarterly* 24:385–405.
32. Hoshino, T., Yura, K., and Hitomi, K. 1995. Optimization analysis for recycle-oriented manufacturing systems. *International Journal of Production Research* 33:855–60.
33. Hu, T., Sheu, J., and Huan, K. 2002. A reverse logistics cost minimization model for the treatment of hazardous wastes. *Transportation Research* 38E:457–73.
34. Imtanavanich, P., and Gupta, S. M. 2004. Multi-criteria decision making for disassembly-to-order system under stochastic yields. In *Proceedings of SPIE—The International Society for Optical Engineering, Environmentally Conscious Manufacturing IV*, vol. 5583, pp. 147–62.
35. Imtanavanich, P., and Gupta, S. M. 2005. Multi-criteria decision making approach in multiple periods for a disassembly-to-order system under product's deterioration and stochastic yields. In *Proceedings of SPIE—The International Society for Optical Engineering, Environmentally Conscious Manufacturing V*, vol. 5997, pp. 599–702.
36. Inderfurth, K. 1997. Simple optimal replenishment and disposal policies for a product recovery system with lead-times. *OR Spectrum* 19:111–22.
37. Jayaraman, V., Guide, Jr., V. D. R., and Srivastava, R. 1999. A closed-loop logistics model for remanufacturing. *Journal of the Operational Research Society* 50:497–508.
38. Kelle, P., and Silver, E. A. 1989. Purchasing policy of new containers considering the random returns of previously issued containers. *IIE Transactions* 21:349–54.
39. Kiesmuller, G. P., and Van der Laan, E. 2001. An inventory model with dependent product demands and returns. *International Journal of Production Economics* 72:73–87.
40. Kongar, E., and Gupta, S. M. 2002. A multi-criteria decision making approach for disassembly-to-order systems. *Journal of Electronics Manufacturing* 11:171–83.
41. Korugan, A., and Gupta, S. M. 1998. A multi-echelon inventory system with returns. *Computers and Industrial Engineering* 35:145–48.

42. Korugan, A., and Gupta, S. M. 2001. An adaptive kanban control mechanism for a single stage hybrid system. In *Proceedings of the International Conference on Environmentally Conscious Manufacturing II*, pp. 175–82.
43. Krikke, H. R., Van Harten, A., and Schuur, P. C. 1999. Business case: Reverse logistic network re-design for copiers. *OR Spectrum* 21:381–409.
44. Krikke, H., Bloemhof-Ruwaard, J., and Van Wassenhove, L. N. 2003. Concurrent product and closed-loop supply chain design with an application to refrigerators. *International Journal of Production Research* 41:3689–719.
45. Krikke, H., Leblanc, I., and van de Velde, S. 2004. Product modularity and the design of closed-loop supply chains. *California Management Review* 46:23–39.
46. Kroon, L., and Vrijens, G. 1995. Returnable containers: An example of reverse logistics. *International Journal of Physical Distribution and Logistics Management* 25:56–68.
47. Krupp, J. 1993. Structuring bills of material for automobile remanufacturing. *Production and Inventory Management Journal* 34:46–52.
48. Lambert, A. J. D., and Gupta, S. M. 2002. Demand-driven disassembly optimization for electronic products. *Journal of Electronics Manufacturing* 11:121–35.
49. Lambert, A. J. D., and Gupta, S. M. 2005. *Disassembly modeling for assembly, maintenance, reuse, and recycling.* Boca Raton, FL: CRC Press.
50. Lambert, A. J. D. 1997. Optimal disassembly of complex products. *International Journal of Production Research* 35:2509–23.
51. Lieckens, K., and Vandaele, N. 2007. Reverse logistics network design with stochastic lead times. *Computers and Operations Research* 34:395–416.
52. Lim, G. H., Kasumastuti, R. D., and Piplani, R. 2005. Designing a reverse supply chain network for product refurbishment. In *Proceedings of the International Conference of Simulation and Modeling.*
53. Listes, O., and Dekker, R. 2005. A stochastic approach to a case study for a product recovery network design. *European Journal of Operational Research* 160:268–87.
54. Listes, O. 2007. A generic stochastic model for supply-and-return network design. *Computers and Operations Research* 34:417–42.
55. Louwers, D., Kip, B. J., Peters, E., Souren, F., and Flapper, S. D. P. 1999. A facility location allocation model for reusing carpet materials. *Computers and Industrial Engineering* 36:855–69.
56. Lu, Z., and Bostel, N. 2007. A facility location model for logistics system including reverse flows: The case of remanufacturing activities. *Computers and Operations Research* 34:299–323.
57. Mabini, M. C., Pintelon, L. M., and Gelders, L. F. 1992. EOQ type formulation for controlling repairable inventories. *International Journal of Production Economics* 28:21–33.
58. Meade, L., Sarkis, J., and Presley, A. 2007. The theory and practice of reverse logistics. *International Journal of Logistics Systems and Management* 3:56–84.
59. Moyer, L. K., and Gupta, S. M. 1997. Environmental concerns and recycling/disassembly efforts in the electronic industry. *Journal of Electronics Manufacturing* 7:1–22.
60. Muckstadt, J., and Isaac, M. H. 1981. An analysis of single item inventory systems with returns. *Naval Research Logistics Quarterly* 28:237–54.
61. Nagurney, A., and Toyasaki, F. 2005. Reverse supply chain management and electronic waste recycling: A multitiered network equilibrium framework for e-cycling. *Transportation Research* 41E:1–28.

62. Panisset, B. 1988. MRP II for repair/refurbish industries. *Production and Inventory Management Journal* 29:12–15.
63. Penev, K. D., and De Ron, A. J. 1996. Determining of disassembly strategy. *International Journal of Production Research* 34:495–506.
64. Pnueli, Y., and Zussman, E. 1997. Evaluating the end-of-life value of a product and improving it by redesign. *International Journal of Production Research* 35:921–42.
65. Ravi, V., Ravi, S., and Tiwari, M. K. 2005. Analyzing alternatives in reverse logistics for end-of-life computers: ANP and balanced scorecard approach. *Computers and Industrial Engineering* 48:327–56.
66. Reimer, B., Sodhi, M. S., and Knight, W. A. 2000. Optimizing electronics end-of-life disposal costs. In *Proceedings of IEEE Symposium on Electronics and the Environment*, pp. 342–47.
67. Richter, K. 1997. Pure and mixed strategies for the EOQ repair and waste disposal problem. *OR Spectrum* 19:123–29.
68. Richter, K. 1996. The EOQ repair and waste disposal model with variable setup numbers. *European Journal of Operational Research* 95:313–24.
69. Richter, K. 1996. The extended EOQ repair and waste disposal model. *International Journal of Production Economics* 45:443–48.
70. Rubio, S., Chamarro, A., and Miranda, J. F. 2008. Characteristics of research on reverse logistics (1999–2005). *International Journal of Production Research* 46:1099–1120.
71. Salema, M. I. G., Barbosa-Povoa, A. P., and Novais, A. Q. 2007. An optimization model for the design of a capacitated multi-product reverse logistics network with uncertainty. *European Journal of Operational Research* 179:1063–77.
72. Salomon, M., van der Laan, E. A., Dekker, R., Thierry, M., and Ridder, A. A. N. 1994. *Product remanufacturing and its effect on production and inventory control.* ERASM Management Report Series 172, Erasmus University, Rotterdam, The Netherlands.
73. Savaskan, R., and Van Wassenhove, L. N. 2006. Reverse channel design: The case of competing retailers. *Management Science* 52:1–14.
74. Savaskan, R. C., Bhattacharya, S., and Van Wassenhove Luk, N. 2004. Closed-loop supply chain models with product remanufacturing. *Management Science* 50:239–52.
75. Simpson, V. P. 1978. Optimum solution structure for a repairable inventory problem. *Operations Research* 26:270–81.
76. Spengler, T., Puchert, H., Penkuhn, T., and Rentz, O. 1997. Environmental integrated production and recycling management. *European Journal of Operations Research* 97:308–26.
77. Teunter, R. H., and Vlachos, D. 2002. On the necessity of a disposal option for returned items that can be remanufactured. *International Journal of Production Economics* 75:257–66.
78. Teunter, R. H. 2001. Economic ordering quantities for remanufacturable item inventory system. *Naval Research Logistics* 48:484–95.
79. Thierry, M., Salomon, M., van Nunen, J., and van Wassenhove, L. N. 1995. Strategic issues in product recovery management. *California Management Review* 37:114–35.
80. Udomsawat, G. and Gupta, S. M. 2008. MultiKanban system for disk assembly line. In S. M. Gupta and A. J. D. Lambert (Eds.), *Environment conscious manufacturing*. Boca Raton, FL: CRC Press..

81. Uzsoy, R. 2007. Production planning for companies with product recovery and remanufacturing capability. In *Proceedings of the IEEE International Symposium on Electronics and the Environment*, pp. 285–90.
82. Van der Laan, E., and Salomon, M. 1997. Production planning and inventory control with remanufacturing and disposal. *European Journal of Operational Research* 102:264–78.
83. Van der Laan, E., Dekker, R., Salomon, M., and Ridder, A. 1996. An (s, Q) inventory model with remanufacturing and disposal. *International Journal of Production Economics* 46:339–50.
84. Veerakamolmal, P., and Gupta, S. M. 1999. Analysis of design efficiency for the disassembly of modular electronic products. *Journal of Electronics Manufacturing* 9:79–95.
85. Vlachos, D., Georgiadis, P., and Iakovou, E. 2007. A systems dynamic model for dynamic capacity planning of remanufacturing in closed-loop supply chains. *Computers and Operations Research* 34:367–94.
86. Wojanowski, R., Verter, V., and Boyaci, T. 2007. Retail-collection network design under deposit-refund. *Computers and Operations Research* 34:324–45.

4

Quantitative Modeling Techniques

4.1 Introduction

In this book, quantitative models for a number of crucial issues in strategic planning of reverse and closed-loop supply chains are presented. Depending on the decision-making situation, each model makes use of one or more of the quantitative techniques introduced in this chapter. It must be noted that only the basic concepts of each technique are presented here.

This chapter is organized as follows: Sections 4.2 through 4.19 introduce the concepts of analytic hierarchy process (including eigen vector method), analytic network process, fuzzy logic, extent analysis method, fuzzy multi-criteria analysis method, quality function deployment, method of total preferences, linear physical programming, goal programming, technique for order preference by similarity to ideal solution (TOPSIS), Borda's choice rule, expert systems, Bayesian updating, Taguchi loss function, Six Sigma, neural networks, geographical information systems, and linear integer programming, respectively. Finally, section 4.20 gives some conclusions.

4.2 Analytic Hierarchy Process and Eigen Vector Method

The analytic hierarchy process (AHP) is a tool supported by simple mathematics that enables decision makers to explicitly weigh tangible and intangible criteria against each other for the purpose of resolving conflict or setting priorities. The process has been formalized by Saaty [1] and is used in a wide variety of problem areas, for example, siting landfills [2], evaluating employee performance [3], and selecting a doctoral program [4].

In a large number of cases (for example, [5]), the tangible and intangible criteria are considered independent of each other; in other words, those criteria do not in turn depend upon subcriteria and so on. The AHP in such cases is conducted in two steps: (1) weigh independent criteria, each of which can compare two or more decision alternatives, using pair-wise judgments, and

(2) compute the relative ranks of decision alternatives using pair-wise judgments with respect to each independent criterion.

1. *Computation of relative weights of criteria*: AHP enables a person to make pair-wise judgments of importance between independent criteria with respect to the scale shown in table 4.1. The resulting matrix of comparative importance values is used to weigh the independent criteria by employing mathematical techniques like eigen value, mean transformation, or row geometric mean. This step is called the eigen vector method if an eigen vector is the employed mathematical technique.

2. *Computation of the relative ranks*: Pair-wise judgments of importance using the scale shown in table 4.1 are computed for the decision alternatives as well. These judgments are obtained with respect to each independent criterion considered in step 1. The resulting matrix of comparative importance values is used to rank the decision alternatives by employing mathematical techniques like eigen value, mean transformation, or row geometric mean.

The degrees of consistency of pair-wise judgments in steps 1 and 2 are measured using an index called the consistency ratio (*CR*). Perfect consistency implies a value of zero for *CR*. However, perfect consistency cannot be demanded because, as human beings, we are often biased and inconsistent in our subjective judgments. Therefore, it is considered acceptable if *CR* is less than or equal to 0.1. For *CR* values greater than 0.1, the pair-wise judgments must be revised before the weights of criteria and the ranks of decision alternatives are computed. *CR* is computed using the formula

$$CR = \frac{(\lambda \max - n)}{(n - 1)(R)} \tag{4.1}$$

where $\lambda\max$ is the principal eigen value of the matrix of comparative importance values, *n* is the number of rows (or columns) in the matrix, and *R* is

TABLE 4.1

Scale for pair-wise judgments

Comparative importance	Definition
1	Equally important
3	Moderately more important
5	Strongly important
7	Very strongly more important
9	Extremely more important
2, 4, 6, 8	Intermediate judgment values

TABLE 4.2

Random index value for each n value

n	1	2	3	4	5	6	7	8	9	10
R	0	0	0.58	0.90	1.12	1.24	0.32	1.41	1.45	1.49

the random index for each n value that is greater than or equal to 1. Table 4.2 shows various R values for n values ranging from 1 to 10.

The AHP is illustrated in the form of a hierarchy of three levels, where the first level contains the primary objective, the second level contains the independent criteria, and the last level contains the decision alternatives. Also, an important feature of the AHP is that the tangible and intangible criteria in the second level must be chosen in such a way that they can somehow help the decision maker in comparing two or more decision alternatives.

4.3 Analytic Network Process

Analytic network process [6] (ANP) generalizes the AHP. AHP assumes independence among the criteria and subcriteria considered in the decision making, but real-life situations warrant against such assumption. ANP allows for dependence within a set of criteria (inner dependence) as well as between sets of criteria (outer dependence); therefore, ANP goes beyond AHP [7]. Whereas AHP assumes a unidirection hierarchical relationship among the decision levels, ANP allows for a more complex relationship among decision levels and attributes, as it does not require a strict hierarchical structure. The looser network structure in ANP allows the representation of any decision problem, irrespective of which criteria come first or which come next. Compared to AHP, ANP requires more calculations and requires a more careful track of the pair-wise judgment matrices. ANP is used in a wide variety of problem areas (for example, analyzing alternatives in reverse logistics for end-of-life computers [7]; evaluating connection types in design for disassembly [8]; and modeling the metrics of lean, agile, and leagile supply chain [9]). The steps involved in the ANP methodology are as follows:

Step 1: *Model development and problem formulation*: In this step, the decision problem is structured into its constituent components. The relevant criteria, the subcriteria, and alternatives are chosen and structured in the form of a control hierarchy.

Step 2: *Pair-wise comparisons*: In this step, the decision maker is asked to carry out a series of pair-wise comparisons with respect to the scale shown in table 4.1, where two main criteria are simultaneously compared with respect to the problem objective, two subcriteria are simultaneously compared with respect to their main criteria, and pair-wise comparisons are performed to address the interdependencies among the subcriteria. The relative matrix of comparative importance values is then used to weigh the criteria using mathematical techniques like eigen vector, mean transformation, or row geometric mean.

Step 3: *Super matrix formulation*: The super matrix allows for a resolution of interdependencies that exist among the subcriteria. It is a partitioned matrix, where each submatrix is composed of a set of relationships between and within the levels as represented by the decision maker's model. The super matrix M is made to converge to obtain a long-term stable set of weights. For convergence, M must be made column stochastic; that is done by raising M to the power of 2^{k+1}, where k is an arbitrarily large number.

Step 4: *Selection of the best alternative*: The selection of the best alternative depends on the desirability index. The desirability index, *DI*, for alternative i is defined as

$$DI = \sum_{j-1}^{J}\sum_{k=1}^{K_j} P_j A_{kj}^D A_{kj}^I S_{ikj} \qquad (4.2)$$

where P_j is the relative importance weight of main criterion j; A_{kj}^D is the relative importance weight for subcriterion k of main criterion j for the dependency (*D*) relationships among subcriteria; A_{kj}^I is the stabilized relative importance weight (determined by the super matrix) for subcriterion k of main criterion j for interdependency (*I*) relationships among subcriteria; and S_{ikj} is the relative impact of alternative i on subcriterion k of main criterion j.

4.4 Fuzzy Logic

Expressions such as "probably so," "not very clear," and "very likely," which are often heard in daily life, carry a touch of imprecision with them. This imprecision or vagueness in human judgments is referred to as *fuzziness* in the scientific literature. As the decision-making problem's intensity grows, this imprecision leads to results that can often be misleading, if the fuzziness is not taken into account. Zadeh [10] first proposed fuzzy logic, after which an increasing number of studies have dealt with fuzziness in problems by applying fuzzy logic.

When dealing with factors with uncertain or imprecise values, people use linguistic values like high, low, good, and medium, to describe those factors. For example, height may be a factor with an imprecise value, so its linguistic value can be "very tall" or "very short." Fuzzy logic is primarily concerned with quantifying vagueness in human perceptions and thoughts. The transition from vagueness to quantification is performed by the application of fuzzy logic, as shown in figure 4.1.

Zadeh proposed a membership function to deal with quantifying vagueness. Each quantified linguistic value is associated with a grade membership value belonging to the interval [0, 1] by means of a membership function. Thus, a fuzzy set can be defined as $\forall x \in X, \mu_A(x) \in [0,1]$, where μ_A is the degree of membership, ranging from 0 to 1, of a quantity x of the linguistic value, A, over the universe of quantified linguistic values, X. X is essentially a set of real numbers. The more x fits A, the larger the degree of membership of x. If a quantity has a degree of membership equal to 1, this reflects a complete fitness between the quantity and the linguistic value. On the other hand, if the degree of membership of a quantity is 0, then that quantity does not belong to the linguistic value. The membership function looks like a typical cumulative probability function; however, the value of a membership function represents the possibility of a fuzzy event, whereas the value of a cumulative probability function represents the cumulative probability of a statistical event.

A triangular fuzzy number (TFN) [11] is a fuzzy set with three parameters (l, m, u), each representing a quantity of a linguistic value associated with a degree of membership of either 0 or 1. Figure 4.2 shows a graphical depiction of a TFN. The parameters l, m, and u denote the smallest possible, most promising, and largest possible quantities that describe the linguistic value.

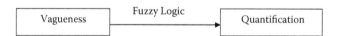

FIGURE 4.1
Application of fuzzy logic.

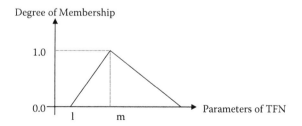

FIGURE 4.2
Triangular fuzzy number.

Each TFN, P, has linear representations on its left- and right-hand side such that its membership function can be defined as

$$\mu_P = \begin{cases} 0, & x < l \\ (x-l)/(m-l), & l \le x \le m \\ (u-x)/(u-m), & m \le x \le u \\ 0, & x > u \end{cases} \tag{4.3}$$

For each quantity x increasing from l to m, the corresponding membership function linearly increases from 0 to 1, and while x increases from m to u, the corresponding membership function decreases linearly from 1 to 0.

The basic operations on TFNs are as follow [11, 12]; for example, $P_1 = (l, m, u)$ and $P_2 = (x, y, z)$:

$$P_1 + P_2 = (l + x, m + y, u + z) \qquad \text{addition} \tag{4.4}$$

$$P_1 - P_2 = (l - z, m - y, u - x) \qquad \text{subtraction} \tag{4.5}$$

$$P_1 \times P_2 = (l \times x, m \times y, u \times z) \qquad \text{multiplication} \tag{4.6}$$

$$\frac{P_1}{P_2} = \left(\frac{l}{z}, \frac{m}{y}, \frac{u}{x} \right) \qquad \text{division} \tag{4.7}$$

Defuzzification is a technique to convert a fuzzy number into a crisp real number. There are several methods available for this purpose. The center-of-area method [13] converts a fuzzy number $P = (l, m, u)$ into a crisp number Q, where

$$Q = \frac{(u - l) + (m - l)}{3} + l \tag{4.8}$$

Defuzzification might be necessary in two situations: (1) when comparison between two fuzzy numbers is difficult to perform, and (2) when a fuzzy number to be operated on has negative parameters (in other words, we make sure that upon performing an arithmetic operation on a TFN, we get a TFN only; for example, squaring the TFN (1, 0, 1) using equation (4.6) leads to (1, 0, 1), which is not a TFN, and hence we defuzzify (–1, 0, 1) before squaring it).

4.5 Extent Analysis Method

Chang [14] proposed a new approach to handle situations that require use of both fuzzy logic and AHP/ANP. First, TFNs (see section 4.4) are used for pair-wise comparisons; then by using extent analysis method [15], the synthetic extent value of the pair-wise comparison is introduced, and by applying the principle of comparison of fuzzy numbers, the weight vectors with respect to each element under a certain criterion can be computed. The steps involved in the methodology are as follows.

Let $X = \{x_1, x_2, ..., x_n\}$ be an object set and $U = \{u_1, u_2, ..., u_m\}$ be a goal set. According to the extent analysis method, each object is taken and an extent analysis for each goal, g_i, is performed. Therefore, m extent analysis values for each object can be obtained, with the following signs: $M^1_{gi}, M^2_{gi}, ..., M^m_{gi}$, $i = 1, 2, ..., n$, where all the M^j_{gi} ($j = 1, 2, ..., m$) are TFNs.

Step 1: The value of fuzzy synthetic extent with respect to the ith object is defined as

$$S_i = \sum_{j=1}^{m} M^j_{gi} \otimes \left[\sum_{i=1}^{n} \sum_{j=1}^{m} M^j_{gi} \right]^{-1}$$

(4.9)

In order to obtain $\sum_{j=1}^{m} M^j_{gi}$, perform the fuzzy addition operation of m extent analysis values for a particular matrix such that

$$\sum_{j=1}^{m} M^j_{gi} = \left(\sum_{j=1}^{m} l_j, \sum_{j=1}^{m} m_j, \sum_{j=1}^{m} u_j \right)$$

(4.10)

To obtain

$$\left[\sum_{i=1}^{n} \sum_{j=1}^{m} M^j_{gi} \right]^{-1},$$

perform the fuzzy addition operation of M^j_{gi} ($j = 1, 2, ..., m$) values such that

$$\sum_{i=1}^{n}\sum_{j=1}^{m}M_{gi}^{j}=\left(\sum_{i=1}^{n}l_i,\sum_{i=1}^{n}m_i,\sum_{i=1}^{n}u_i\right)$$

(4.11)

and then compute the inverse of the vector.

Step 2: The degree of possibility of $M_2 = (l_2, m_2, u_2) \geq M_1 = (l_1, m_1, u_1)$ is expressed as

$$V(M_2 \geq M_1)$$
$$= \text{hgt}(M_1 \geq M_2)$$
$$= \{1, \text{ if } m_2 \geq m_1; 0, \text{ if } l_1 \geq u_2; (l_1 - u_2)/((m_2 - u_2) - (m_1 - l_1))\} \quad (4.12)$$

To compare M_1 and M_2, both $V(M_2 \geq M_1)$ and $V(M_1 \geq M_2)$ are required.

Step 3: The degree of possibility for a convex fuzzy number to be greater than k convex fuzzy numbers M_i ($i = 1, 2, \ldots, k$) can be defined as

$$V(M \geq M_1, M_2, M_k)$$
$$= V[(M \geq M_1) \text{ and } (M \geq M_2) \text{ and } \ldots \text{ and } (M \geq M_k)]$$
$$= \min V(M \geq M_i), i = 1, 2, k \quad (4.13)$$

Let $d'(A_i) = \min V(S_i \geq S_k)$, for $k = 1, 2, n; k \neq i$. Then the weight vector is given by

$$W' = (d'(A_1), d'(A_2), d'(A_n))^T \quad (4.14)$$

Step 4: The weight vector obtained in step 3 is normalized to get the normalized weights.

These steps are applied to deduce the weights of main criteria with respect to the goal, subcriteria with respect to the main criteria, and alternatives with respect to the main and subcriteria. The rest of the procedure is similar to the traditional AHP/ANP.

4.6 Fuzzy Multicriteria Analysis Method

Multicriteria analysis problems require the decision maker to make qualitative assessments regarding the performance of the decision alternatives with respect to each independent criterion and the relative importance of each independent criterion with respect to the overall objective of the problem. As a result, uncertain subjective data are present that make the decision-making process complex [16]. AHP enables a person to make pair-wise judgments of importance between the independent criteria as well as the decision alternatives. However, traditional AHP is criticized for its unbalanced scale of judgment and failure to precisely handle the inherent uncertainty and vagueness in carrying out pair-wise comparisons.

Deng [17] proposed a multicriteria analysis approach that extends Saaty's AHP (see section 4.2) to deal with the imprecision and subjectiveness in the pair-wise comparisons. TFNs (see section 4.4) are used for pair-wise comparisons, and the concept of extent analysis method (see section 4.5) is applied to solve the fuzzy reciprocal matrix for determining the criteria importance and alternative performance. The α-cut concept is used to transform the fuzzy performance matrix representing the overall performance of all alternatives with respect to each criterion into an interval performance matrix. An overall performance index for each alternative across all criteria that incorporates the decision maker's attitude toward risk is obtained by applying the concept of similarity to the ideal solution [18] using the vector-matching function.

The selection process starts with determining the criteria of importance and performance of alternatives. By using TFNs, a fuzzy reciprocal matrix for criteria importance (*W*) or alternative performance with respect to a specific criterion (*C$_j$*) can be determined as

$$W \text{ or } C_j = \begin{bmatrix} \overline{a}_{11} & \overline{a}_{12} & \cdots & \overline{a}_{1k} \\ \overline{a}_{21} & \overline{a}_{22} & \cdots & \overline{a}_{2k} \\ \cdots & \cdots & \cdot\cdot & \cdots \\ \overline{a}_{k1} & \overline{a}_{k2} & \cdots & \overline{a}_{kk} \end{bmatrix}$$

(4.15)

where

$$\overline{a}_{ls} = \begin{cases} \overline{1}, \overline{3}, \overline{5}, \overline{9}, l < s, \\ 1, l = s, l, s = 1, 2, ...k; k = m, or, n, \\ 1/\overline{a}_{sl}, l > s \end{cases}$$

(4.16)

By applying the extent analysis method, the corresponding criteria weights (*w$_j$*) or alternative performance ratings (*x$_{ij}$*) with respect to a specific criterion *C$_j$* can be determined as

$$x_{ij} \text{ or } w_j = \sum_{s=1}^{k} \overline{a}_{ls} \div \sum_{l=1}^{k} \sum_{s=1}^{k} \overline{a}_{ls}$$

(4.17)

where $i = 1, 2, ..., n$; $j = 1, 2, ..., m$; and $k = m$ or n depending on whether the reciprocal judgment matrix is for assessing the performance ratings of alternatives or weights of criteria involved. The decision matrix (X) and the weight vector (W) can be respectively determined as

$$X = \begin{bmatrix} x_{11} & x_{12} & \cdots & x_{1m} \\ x_{21} & x_{22} & \cdots & x_{2m} \\ \cdots & \cdots & \cdots & \cdots \\ x_{n1} & x_{n2} & \cdots & x_{nm} \end{bmatrix}$$

(4.18)

$$W = (w_1, w_2, ..., w_m)$$

(4.19)

where x_{ij} represents the resultant fuzzy performance assessment of alternative A_i ($i = 1, 2, n$) with respect to criterion C_j, and w_j is the resultant fuzzy weight of criterion C_j ($j = 1, 2, ..., m$) with respect to the overall goal of the problem. A fuzzy performance matrix Z representing the overall performance of all alternatives with respect to each criterion is obtained by multiplying the weight vector by the decision matrix.

$$Z = \begin{bmatrix} w_1 x_{11} & w_2 x_{12} & \cdots & w_m x_{1m} \\ w_1 x_{21} & w_2 x_{22} & \cdots & w_m x_{2m} \\ \cdots & \cdots & \cdots & \cdots \\ w_1 x_{n1} & w_2 x_{n2} & \cdots & w_m x_{nm} \end{bmatrix}$$

(4.20)

An interval performance matrix [16] is derived by using an α-cut on the performance matrix, where $0 \leq \alpha \leq 1$. The value of α represents the decision maker's degree of confidence in his or her fuzzy assessments regarding the alternative ratings and criteria weights. The larger the value of α, the more confident the decision maker is about the fuzzy assessments, viz., the assessments are closer to the most possible value a_2 of the triangular fuzzy number (a_1, a_2, a_3).

$$Z_\alpha = \begin{bmatrix} \left[z_{11l}^\alpha, z_{11r}^\alpha\right] & \left[z_{12l}^\alpha, z_{12r}^\alpha\right] & \cdots & \left[z_{1ml}^\alpha, z_{1mr}^\varepsilon\right] \\ \cdots & \cdots & \cdots & \cdots \\ \cdots & \cdots & \cdots & \cdots \\ \left[z_{n1l}^\alpha, z_{n1r}^\alpha\right] & \left[z_{n2l}^\alpha, z_{n2r}^\alpha\right] & \cdots & \left[z_{nml}^\alpha, z_{nmr}^\alpha\right] \end{bmatrix}$$

(4.21)

An overall crisp performance matrix that incorporates the decision maker's attitude toward risk, using an optimism index λ ($\lambda = 1$ implies the decision maker has an optimistic view, 0 implies a pessimistic view, and 0.5 implies a moderate view), is calculated.

$$
z_\alpha^{\lambda'} = \begin{bmatrix} z_{11\alpha}^{\lambda'} & z_{12\alpha}^{\lambda'} & \cdots & z_{1m\alpha}^{\lambda'} \\ \cdots & \cdots & \cdots & \cdots \\ \cdots & \cdots & \cdots & \cdots \\ z_{n1\alpha}^{\lambda'} & z_{n2\alpha}^{\lambda'} & \cdots & z_{nm\alpha}^{\lambda'} \end{bmatrix}
$$

(4.22)

where

$$
z_{ij\alpha}^{\lambda'} = \lambda z_{ijr}^{\alpha} + (1-\lambda)z_{ijl}^{\alpha}, \lambda \in [0,1]
$$

(4.23)

A normalized performance matrix with respect to each criterion is calculated from equation (4.22).

$$
Z_\alpha = \begin{bmatrix} z_{11\alpha}^{\lambda} & z_{12\alpha}^{\lambda} & \cdots & z_{1m\alpha}^{\lambda} \\ z_{21\alpha}^{\lambda} & z_{22\alpha}^{\lambda} & \cdots & z_{2m\alpha}^{\lambda} \\ \cdots & \cdots & \cdots & \cdots \\ z_{n1\alpha}^{\lambda} & z_{n2\alpha}^{\lambda} & \cdots & z_{nm\alpha}^{\lambda} \end{bmatrix}
$$

(4.24)

where

$$
z_{ij\alpha}^{\lambda} = z_{ij\alpha}^{\lambda'} \div \sqrt{\sum_{i=1}^{n}(z_{ij\alpha}^{\lambda'})^2}
$$

(4.25)

Zeleny [18] introduced the concept of ideal solution in multiattribute decision analysis that was further extended by Hwang and Yoon [19], including negative solution to avoid the worst decision outcome. In line with this concept, the positive- and negative-ideal solutions, respectively, can be determined by selecting maximum and minimum values across all alternatives with respect to each criterion as follows:

$$
A_\alpha^{\lambda+} = (z_{1\alpha}^{\lambda+}, z_{2\alpha}^{\lambda+}, ..., z_{m\alpha}^{\lambda+})
$$
$$
A_\alpha^{\lambda-} = (z_{1\alpha}^{\lambda-}, z_{2\alpha}^{\lambda-}, ..., z_{m\alpha}^{\lambda-})
$$

(4.26)

where

$$z_{j\alpha}^{\lambda+} = \max(z_{j\alpha}^{\lambda}, z_{2j\alpha}^{\lambda}, ..., z_{nj\alpha}^{\lambda})$$

$$z_{j\alpha}^{\lambda-} = \min(z_{j\alpha}^{\lambda}, z_{2j\alpha}^{\lambda}, ..., z_{nj\alpha}^{\lambda}) \tag{4.27}$$

By applying the vector-matching function, the degree of similarity between each alternative and the positive- and negative-ideal solutions can be calculated as

$$S_{i\alpha}^{\lambda+} = A_{i\alpha}^{\lambda} A_{i\alpha}^{\lambda+} / \max(A_{i\alpha}^{\lambda} A_{i\alpha}^{\lambda}, A_{\alpha}^{\lambda+} A_{\alpha}^{\lambda+})$$

$$S_{i\alpha}^{\lambda-} = A_{i\alpha}^{\lambda} A_{i\alpha}^{\lambda-} / \max(A_{i\alpha}^{\lambda} A_{i\alpha}^{\lambda}, A_{\alpha}^{\lambda-} A_{\alpha}^{\lambda-}) \tag{4.28}$$

where $A_{i\alpha}^{\lambda} = (z_{i1\alpha}^{\lambda}, z_{i2\alpha}^{\lambda}, ..., z_{im\alpha}^{\lambda})$ is the ith row of the overall performance matrix, which represents the corresponding performance of alternative A_i with respect to criterion C_j. The larger the value of $S_{i\alpha}^{\lambda+}, S_{i\alpha}^{\lambda-}$, the higher the degree of similarity between each alternative and the positive-ideal and negative-ideal solutions [20]. A preferred alternative should have a higher degree of similarity to the positive-ideal solution and a lower degree of similarity to the negative-ideal solution. Hence, an overall performance index for each alternative with the decision maker's α level of confidence and λ degree of optimism toward risk can be determined as

$$P_{\alpha i}^{\lambda} = S_{i\alpha}^{\lambda+} / (S_{i\alpha}^{\lambda} + S_{i\alpha}^{\lambda-}), i - 1, 2, ..., n \tag{4.29}$$

The larger the performance index, the most preferred the alternative is.

4.7   Quality Function Deployment

Erol and Ferrell [21] define *performance aspects* as the features that the decision maker wishes to consider in the selection process and *enablers* as the characteristics possessed by the alternatives, which can be used to satisfy the performance aspects.

The absolute technical importance ratings (ATIRs), which measure how effectively each enabler can satisfy all of the performance aspects, are computed by

$$\text{ATIR}_j = \sum_{i=1}^{I} d_i R_{ij} \quad \forall \, j = 1, ..., J \tag{4.30}$$

where d_i is the importance value of performance aspect i relative to the other performance aspects, and R_{ij} is the relationship score for performance aspect i and enabler j. Because there is an ATIR for each enabler j, for the comparison of all enablers, it is normalized to form the relative technical importance rating (RTIR$_j$) as follows:

$$RTIR_j = \frac{ATIR_j}{\sum_{j=1}^{J} ATIR_j} \quad \forall\, j = 1, ..., J$$

(4.31)

4.8 Method of Total Preferences

RTIRs (see section 4.7), together with additional human expert opinions, are used to develop a single measure that reflects the rating of each alternative as follows [21]:

$$TUP_n = \sum_{j=1}^{J} RTIR_j WA_{nj} \quad \forall\, n$$

(4.32)

where TUP_n is the total user preference for alternative n, and WA_{nj} is the (defuzzified) degree to which alternative n can deliver enabler j.

For the purpose of comparison of all alternatives, TUP of each alternative is then normalized as follows:

$$NTUP_n = \frac{TUP_n}{\sum_{n=1}^{N} TUP_n} \quad \forall\, n$$

(4.33)

where $NTUP_n$ is the normalized total preference for alternative n, and N is the total number of alternatives.

The alternative with the highest NTUP is considered the one with the highest potential.

4.9 Linear Physical Programming

In the linear physical programming (LPP) method [22], four distinct classes (1S, 2S, 3S, and 4S) are used to allow the decision maker to express his or her preferences for the value of each criterion (for decision making) in a more detailed, quantitative, and qualitative way than when using a weight-based method like analytic hierarchy process (see section 4.2). These classes are defined as follows: smaller is better (1S), larger is better (2S), value is better (3S), and range is better (4S). Figure 4.3 depicts these different classes.

The value of the pth criterion, g_p, for evaluating the alternative of interest is categorized according to the preference ranges shown on the horizontal axis. Consider, for example, the case of class 1S. The preference ranges are:

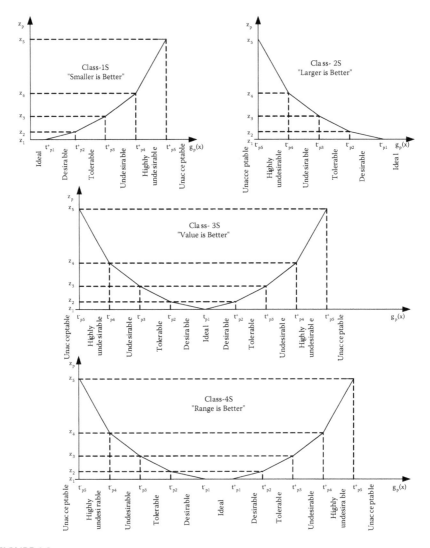

FIGURE 4.3
Soft class functions for linear physical programming.

Ideal range: $g_p \leq t_{p1}^+$

Desirable range: $t_{p1}^+ \leq g_p \leq t_{p2}^+$

Tolerable range: $t_{p2}^+ \leq g_p \leq t_{p3}^+$

Undesirable range: $t_{p3}^+ \leq g_p \leq t_{p4}^+$

Highly undesirable range: $t_{p4}^+ \leq g_p \leq t_{p5}^+$

Unacceptable range: $g_p \geq t_{p5}^+$

The quantities t_{p1}^+ through t_{p5}^+ represent the physically meaningful values that quantify the preferences associated with the pth generic criterion. Consider, for example, the cost criterion for class 1S. The decision maker could specify a preference vector by identifying t_{p1}^+ through t_{p5}^+ in dollars as (10 20 30 40 50). Thus, an alternative having a cost of $15 would lie in the desirable range, an alternative with a cost of $45 would lie in the highly undesirable range, and so on. We can accomplish this for a nonnumerical criterion such as color as well by (1) specifying a numerical preference structure and (2) quantitatively assigning each alternative a specific criterion value from within a preference range (e.g., desirable, tolerable).

The class function, Z_p, on the vertical axis in figure 4.3 is used to map the criterion value, g_p, into a real, positive, and dimensionless parameter (Z_p is, in fact, a piecewise linear function of g_p). Such a mapping ensures that different criteria values, with different physical meanings, are mapped to a common scale. Consider class 1S again. If the value of a criterion, g_p, is in the ideal range, then the value of the class function is small (in fact, zero), whereas if the value of the criterion is greater than t_{p5}^+, that is, in the unacceptable range, then the value of the class function is very high. Class functions have several important properties, including (1) that they are nonnegative, continuous, piecewise linear, and convex, and (2) that the value of the class function, Z_p, at a given range intersection (say, desirable–tolerable) is the same for all class types.

Basically, ranking of the alternatives is performed in four steps, as follows [23]:

Step 1: *Identify criteria for evaluating each of the alternatives.*

Step 2: *Specify preferences for each criterion, based on one of the four classes* (see figure 4.3).

Step 3: *Calculate incremental weights*: Based on the preference structures for the different criteria, the LPP weight algorithm [22] determines incremental weights, Δw_{pr}^+ and Δw_{pr}^- (used in step 4), that represent the incremental slopes of the class functions, Z_p. Here, r denotes the range intersection.

Step 4: *Calculate total score for each alternative*: The formula for the total score, *J*, of the alternative of interest is constructed as a weighted sum of deviations over all ranges ($r = 2$ to 5) and criteria ($p = 1$ to P), as follows:

$$J = \sum_{p=1}^{P}\sum_{r=2}^{5}(\Delta w_{pr}^{-}d_{pr}^{-} + \Delta w_{pr}^{+}d_{pr}^{+})$$

(4.34)

where P represents the total number of criteria (each belonging to one of the four classes in figure 4.3), Δw_{pr}^{+} and Δw_{pr}^{-} are the incremental weights for the *p*th criterion, and d_{pr}^{+} and d_{pr}^{-} represent the deviations of the *p*th criterion value of the alternative of interest from the corresponding target values. An alternative with a lower total score is more desirable than one with a higher total score.

The most significant advantage of using LPP is that no weights need to be specified for the criteria for evaluation. The decision maker only needs to specify a preference structure for each criterion, which has more physical meaning than a physically meaningless weight that is arbitrarily assigned to the criterion.

Note that there are no decision variables in the above ranking procedure. LPP can be used in a problem consisting of decision variables as well, by minimizing *J* in equation (4.34) and subjecting (if necessary) each criterion, g_p, to a constraint that falls into either one of the four classes (also called *soft* classes) in figure 4.3 or one of the following four *hard* classes:

Class 1H: Must be smaller, i.e., $g_p \le t_{p,max}$

Class 2H: Must be larger, i.e., $g_p \ge t_{p,min}$

Class 3H: Must be equal, i.e., $g_p = t_{p,val}$

Class 4H: Must be in range, i.e., $t_{p,min} \le g_p \le t_{p,max}$

4.10   Goal Programming

Linear programming [24] assumes that the objectives of an organization can be encompassed within a single objective function, such as maximizing the total profit or minimizing the total cost. However, this assumption is not always realistic, and there are several cases where the management focuses on a variety of objectives simultaneously. Goal programming provides a way of tackling such situations.

Goal programming (GP; see [25]), generally applied to linear problems, deals with the achievement of specific targets/goals. The basic approach involves formulating an objective function for each objective and seeks a solution that minimizes the sum (weighted sum in case of fuzzy goal programming) of the deviations of these objective functions from their respective goals. To this end, several criteria are to be considered in the problem situation on hand. For each criterion, a target value is determined. Next, the deviation variables are introduced, which may be positive or negative (represented by ρ_k and η_k, respectively). The negative deviation variable, η_k, represents the underachievement of the kth goal. Similarly, the positive deviation variable, ρ_k, represents the overachievement of the kth goal. Finally, for each criterion, the desire to overachieve (minimize η_k) or underachieve (minimize ρ_k) or to satisfy the target value exactly (minimize $\rho_k + \eta_k$) is articulated [26].

Goal programming problems can be categorized according to the type of mathematical programming model. Another categorization is according to how the goals compare in importance. In the case of preemptive goal programming, there is a hierarchy of priority levels for the goals, so the goals of primary importance receive first attention and so forth. In case of nonpreemptive goal programming, all the goals are of roughly comparable importance [24].

In goal programming, it is necessary to specify aspiration levels for the goals, and the overall deviation from the aspiration levels is minimized. In most real-world scenarios, the aspiration levels and weights/importance levels of goals are imprecise in nature. In such situations, fuzzy GP comes in handy, allowing the decision maker to obtain compromising results for multiple goals with varying aspiration levels. In fuzzy GP, the aspiration levels are either in the "more is better" form or "less is better" form. [27]. A linear membership function μ_i that represents goal fuzziness for the "more is better" form is expressed as [28]

$$
\mu_i = \begin{cases} 1 & \text{if } G_i(X) \geq g_i \\ \dfrac{G_i(X) - L_i}{g_i - L_i} & \text{if } L_i \leq G_i(X) \leq g_i \\ 0 & \text{if } G_i(X) \leq L_i \end{cases}
$$

(4.35)

while a linear membership fraction μ_i that represents goal fuzziness for the "less is better" form is expressed as

$$\mu_i = \begin{cases} 1 & \text{if } G_i(X) \leq g_i \\ \dfrac{U_i - G_i(X)}{U_i - g_i} & \text{if } g_i \leq G_i(X) \leq U_i \\ 0 & \text{if } G_i(X) \geq U_i \end{cases} \tag{4.36}$$

where g_i, L_i, and U_i are the aspiration level, lower tolerance limit, and upper tolerrance limit respectively for the fuzzy goal G_i (X).

The simple form of the fuzzy GP problem with *p* fuzzy goals can be stated as:

$$\text{Maximize} \quad V(\mu) = \sum_{i=1}^{p} \mu_i \tag{4.37}$$

$$\text{subject to } \mu_i = \frac{G_i(X) - L_i}{g_i - L_i} \text{ or } \mu_i = \frac{U_i - G_i(X)}{U_i - g_i} \tag{4.38}$$

$$AX \leq b \tag{4.39}$$

$$\mu_i \leq 1 \tag{4.40}$$

$$X, \mu_i \geq 0 \tag{4.41}$$

where $V(\mu)$ is the fuzzy achievement function or fuzzy decision function. The objective is to obtain the μ_i value as close to 1 as possible. The weighted additive model is widely used in GP and multiobjective optimization problems to reflect the relative importance of goals. In this approach, the decision maker assigns weights as coefficients of individual terms in the simple additive fuzzy achievement function to reflect their relative importance. The objective function for the weighted additive model is expressed as

$$V(\mu) = \sum_{i=1}^{p} w_i \mu_i$$
$$\text{Maximize} \tag{4.42}$$

where w_i is the relative weight of the *i*th fuzzy goal. Fibonacci numbers are used to assign weights to the goals. Fibonacci numbers are 0, 1, 1, 2, 3, 5, 8, etc. (the next number is a result of the summation of the previous two numbers).

The concept is applied by starting with numbers 1 and 2. For example, for two goals, the weights would be in the ratio of 1:2, which approximately are 0.33 and 0.66. These weights are assigned to the two goals according to their priority levels.

In some situations, the goals/objectives are not commensurable, or the goals are such that unless a particular goal or subset of goals is achieved, other goals should not be considered. In such situations, the weighting scheme is not appropriate. The problem is divided into k subproblems, where k is the number of priority levels. In the first subproblem, the fuzzy goals belonging to the first priority level will be considered and solved using the simple additive model. At other priority levels, the membership values achieved at earlier priority levels are added as additional constraints. In general, the ith subproblem becomes:

$$\text{Maximize} \sum_s (\mu_s)_{pi} \tag{4.43}$$

$$\text{subject to } \mu_s = \frac{G_s - L_s}{g_s - L_s} \tag{4.44}$$

$$AX \le b \tag{4.45}$$

$$(\mu)_{pr} = (\mu^*)_{pr}, r = 1, 2, \ldots, j-1 \tag{4.46}$$

$$\mu_s \le 1 \tag{4.47}$$

$$X, \mu_i \ge 0, \quad i = 1, 2, \ldots, p \tag{4.48}$$

where $(\mu_s)_{pi}$ refers to the membership functions of the goals in the ith priority level and $(\mu^*)_{pr}$ is the achieved membership function value in the rth ($r \le j - 1$) priority level.

4.11 Technique for Order Preference by Similarity to Ideal Solution (TOPSIS)

The basic concept of the TOPSIS method [19] is that the rating of the alternative selected as the best from a set of different alternatives should have the shortest distance from the ideal solution and the greatest distance from the negative-ideal solution in a geometrical (i.e., Euclidean) sense.

The TOPSIS method evaluates the following decision matrix, which refers to m alternatives that are evaluated in terms of n criteria [29]:

				Criteria		
	C_1	C_2	C_3	...	C_n
Alternatives	w_1	w_2	w_3	...	w_n
A_1	z_{11}	z_{12}	z_{13}	...	z_{1n}
A_2	z_{21}	z_{22}	z_{23}	...	z_{2n}
A_3	z_{31}	z_{32}	z_{33}	...	z_{3n}
.	...				
.	...				
.	...				
.	...				
A_m	z_{m1}	z_{m2}	z_{m3}	...	z_{mn}

where A_i is the ith alternative, C_j is the jth criterion, w_j is the weight (importance value) assigned to the jth criterion, and z_{ij} is the rating (for example, on a scale of 1–10, the higher the rating, the better it is) of the ith alternative in terms of the jth criterion.

The following steps are performed:

Step 1: *Construct the normalized decision matrix.* This step converts the various dimensional measures of performance into nondimensional attributes. An element r_{ij} of the normalized decision matrix R is calculated as follows:

$$r_{ij} = \frac{z_{ij}}{\sqrt{\sum_{i=1}^{m} z_{ij}^2}}$$

(4.49)

Step 2: *Construct the weighted normalized decision matrix.* A set of weights $W = (w_1, w_2, ..., w_n)$ (such that $\Sigma w_j = 1$), specified by the decision maker, is used in conjunction with the normalized decision matrix R to determine the weighted normalized matrix V defined by $V = (v_{ij}) = (r_{ij}w_j)$.

Step 3: *Determine the ideal and the negative-ideal solutions.* The ideal (A^*) and the negative-ideal ($A–$) solutions are defined as follows:

$$A^* = \left\{ \max_i v_{ij} \quad \text{for } i = 1, 2, 3, ..., m \right\}$$

$$= \{p_1, p_2, p_3, ..., p_n\}$$

(4.50)

$$A- = \left\{ \min_i v_{ij} \quad \text{for } i = 1, 2, 3, \ldots, m \right\}$$

$$= \{q_1, q_2, q_3, \ldots, q_n\} \tag{4.51}$$

With respect to each criterion, the decision maker desires to choose the alternative with the maximum rating (it is important to note that this choice varies with the way he or she awards ratings to the alternatives). Obviously, A^* indicates the most preferable (ideal) solution. Similarly, $A-$ indicates the least preferable (negative-ideal) solution.

Step 4: *Calculate the separation distances.* In this step, the concept of the n-dimensional Euclidean distance is used to measure the separation distances of the rating of each alternative from the ideal solution and the negative-ideal solution. The corresponding formulae are

$$S_{i*} = \sqrt{\sum (v_{ij} - p_j)^2} \quad \text{for } i = 1, 2, 3, \ldots, m \tag{4.52}$$

where S_{i*} is the separation (in the Euclidean sense) of the rating of alternative i from the ideal solution, and

$$S_{i-} = \sqrt{\sum (v_{ij} - q_j)^2} \quad \text{for } i = 1, 2, 3, \ldots, m \tag{4.53}$$

where S_{i-} is the separation (in the Euclidean sense) of the rating of alternative i from the negative-ideal solution.

Step 5: *Calculate the relative coefficient.* The relative closeness coefficient for alternative A_i with respect to the ideal solution A^* is defined as follows:

$$C_{i*} = \frac{S_{i-}}{S_{i*} + S_{i-}} \tag{4.54}$$

Step 6: *Rank the preference order.* The best alternative can now be decided according to the preference order of C_{i*}. It is the one with the rating that has the shortest distance to the ideal solution. The way the alternatives are processed in the previous steps reveals that if an alternative has the rating with the shortest distance to the ideal solution, then that rating is guaranteed to have the longest distance to the negative-ideal solution. That means the higher the C_{i*}, the better the alternative.

4.12   Borda's Choice Rule

Borda proposed a method in which marks of $m-1, m-2, ..., 1, 0$ are assigned to the best, second-best, ..., worst alternatives, for each decision maker [30]. That means that a larger mark corresponds to greater preference. The Borda score (maximized consensus mark) for each alternative is then determined as the sum of the individual marks for that alternative, and the alternative with the highest Borda score is declared the winner. That means that the different decision makers unanimously choose the alternative that obtains the largest Borda score as the most preferred one.

4.13   Expert Systems

Expert systems are computer programs that can represent human expertise (knowledge) in a particular domain (area of expertise) and then use a reasoning mechanism (applying logical deduction and induction processes) to manipulate this knowledge in order to provide advice in this domain. Although conventional computer programs also contain knowledge, their main function is to retrieve information and carry out statistical analysis and numerical calculations. They do not reason with this knowledge or make inferences as to what actions to take or conclusions to reach. Thus, what mainly distinguishes expert systems from conventional programs is the capability to reason with knowledge. The main components of an expert system are the following [31]:

- *Knowledge base*: This is where the knowledge is stored. Typically, this consists of a set of rules of the form: if EVIDENCE, then HYPOTHESIS. The knowledge is written in the knowledge base using the syntax of what is termed the *knowledge representation language* (e.g., Lisp and Prolog) of the system.

- *Inference engine*: This reasons with the knowledge resident in the knowledge base using certain mechanisms.

- *Reasoning mechanism*: This traces the path or the knowledge steps used to arrive at a conclusion and can relay it back to the user as the justification for this conclusion. Examples of this mechanism are deduction (cause + rule → effect), abduction (effect + rule → cause), and induction (cause + effect → rule).

- *Uncertainty modeling process*: This aids the inference engine when dealing with uncertainty.

A *shell* is an expert system that is complete except for the knowledge base [31]. Thus, a shell includes an inference engine, a user interface for programming, and a user interface for running the system. Typically, the programming interface comprises a specialized editor for creating rules in a predetermined format and some debugging tools. The user of the shell enters rules in a declarative fashion (if *X*, then *Y*) and ideally should not need to be concerned with the working of the inference engine. Expert system shells are easy to use and allow a simple expert system to be constructed quickly.

4.14 Bayesian Updating

Bayesian updating [32] is an uncertainty modeling technique that assumes that it is possible for an expert in a domain to guess a probability to every hypothesis or assertion in that domain and that this probability can be updated in light of evidence for or against the hypothesis or assertion.

Suppose the probability of a hypothesis H is P(H). Then the formula for the odds of that hypothesis, O(H), is given by

$$O(H) = \frac{P(H)}{1 - P(H)} \tag{4.55}$$

A hypothesis that is absolutely certain, i.e., has a probability of 1, has infinite odds. In practice, limits are often set on odds values so that, for example, if O(H) > 1,000, then H is true, and if O(H) < 0.01, then H is false.

The standard formula for updating the odds of hypothesis H, given that evidence E is observed, is

$$O(H|E) = (A).O(H) \tag{4.56}$$

where O(H|E) is the odds of H, given the presence of evidence E, and A is the *affirms* weight of E. The definition of A is

$$A = \frac{P(E|H)}{P(E|{\sim}H)} \tag{4.57}$$

where P(E|H) is the probability of E, given that H is true, and P(E|~H) is the probability of E, not given that H is true.

Bayesian updating assumes that the absence of supporting evidence is equivalent to the presence of opposing evidence. The standard formula for updating the odds of a hypothesis H, given that the evidence E is absent, is

$$O(H|\sim E) = (D).O(H) \qquad (4.58)$$

where $O(H|\sim E)$ is the odds of H, given the absence of evidence E, and D is the *denies* weight of E. The definition of D is

$$D = \frac{P(\sim E|H)}{P(\sim E|\sim H)} = \frac{1-P(E|H)}{1-P(E|\sim H)} \qquad (4.59)$$

If a given piece of evidence E has an *affirms* weight A that is greater than 1, then its *denies* weight must be less than 1, and vice versa. Also, if A > 1 and D < 1, then the presence of evidence E is supportive of hypothesis H. Similarly, if A < 1 and D > 1, then the absence of E is supportive of H.

For example, while controlling a power station boiler, a rule "IF (temperature is high) and NOT (water level is low) THEN (pressure is high)" can also be written as "IF (temperature is high—AFFIRMS A_1, DENIES D_1) AND (water level is low—AFFIRMS A_2, DENIES D_2) THEN (pressure is high)." Here,

$$A_1 = \frac{P(\text{Temperature is high} \mid \text{Pressure is high})}{P(\text{Temperature is high} \mid \sim\text{Pressure is high})}$$

$$D_1 = \frac{P(\sim\text{Temperature is high} \mid \text{Pressure is high})}{P(\sim\text{Temperature is high} \mid \sim\text{Pressure is high})}$$

$$A_2 = \frac{P(\text{Water level is low} \mid \text{Pressure is high})}{P(\text{Water level is low} \mid \sim\text{Pressure is high})}$$

$$D_2 = \frac{P(\sim\text{Temperature is high} \mid \text{Pressure is high})}{P(\sim\text{Temperature is high} \mid \sim\text{Pressure is high})}$$

Sometimes, evidence is neither definitely present nor definitely absent. For example, if one is diagnosing a TV set that is not functioning properly, it is not definite if this is due to a malfunctioning picture tube. In such a case, depending upon the value of the probability of the evidence P(E), the affirms and denies weights are modified using the following formulae:

$$A' = [2.(A-1).P(E)]+2-A \qquad (4.60)$$

$$D' = [2.(1-D).P(E)]+D \qquad (4.61)$$

When P(E) is greater than 0.5, the *affirms* weight is used to calculate $O(H|E)$, and when P(E) is less than 0.5, the *denies* weight is used.

If n statistically independent pieces of evidence are found that support or oppose a hypothesis H, then the updating equations are given by

$$O(H|E_1 \& E_2 \& E_3......E_n) = (A_1).(A_2).(A_3).......(A_n).O(H) \qquad (4.62)$$

and

$$O(H|\sim E_1 \& \sim E_2 \& \sim E_3... \sim E_n) = (D_1).(D_2).(D_3).......(D_n).O(H) \qquad (4.63)$$

A_i and D_i are given by equations (4.64) and (4.65), respectively.

$$A_i = \frac{P(E_i|H)}{P(E_i|\sim H)} \qquad (4.64)$$

$$D_i = \frac{P(\sim E_i|H)}{P(\sim E_i|\sim H)} \qquad (4.65)$$

4.15 Taguchi Loss Function

In traditional systems, the product is accepted if the product measurement falls within the specification limits [33]. Otherwise, the product is rejected. The quality losses occur only when the product deviates beyond the specification limits, thereby becoming unacceptable. These costs tend to be constant and relate to the costs of bringing the product back into the specification range. Taguchi [33] suggests a narrower view of characteristic acceptability by indicating that any deviation from a characteristic's target value results in a loss. If a characteristic measurement is the same as the target value, the loss is zero. Otherwise, the loss can be measured using a quadratic function, after which actions are taken to reduce systematically the variation from the target value.

There are three types of Taguchi loss functions: "target is best" (see figure 4.4), "smaller is better" (see figure 4.5), and "larger is better" (see figure 4.6).

If $L(y)$ is the loss associated with a particular value of characteristic y, m is the target value of the specification, and k is the loss coefficient whose value is constant depending on the cost at the specification limits and width of the specification, for the "target is best" type,

$$L(y) = k(y-m)^2 \qquad (4.66)$$

For the "smaller is better" type,

$$L(y) = k(y)^2 \qquad (4.67)$$

For the "larger is better" type,

$$L(y) = k/(y)^2 \qquad (4.68)$$

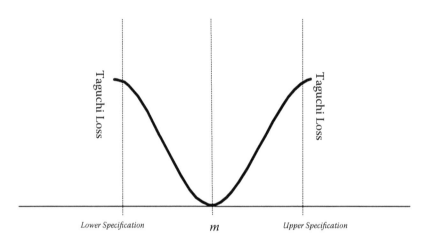

FIGURE 4.4
"Target is best" Taguchi loss.

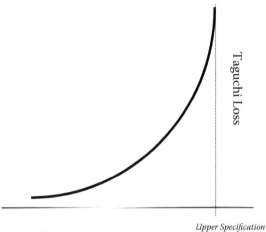

FIGURE 4.5
"Smaller is better" Taguchi loss.

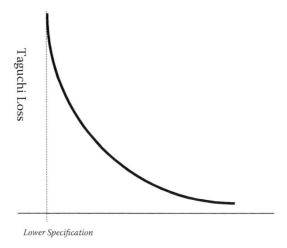

FIGURE 4.6
"Larger is better" Taguchi loss.

4.16 Six Sigma

Statistical process control techniques help managers achieve and maintain a process distribution that does not change in terms of its mean and variance [34]. The control limits on the control charts signal when the mean or variability of the process changes. However, a process that is in statistical control may not be producing outputs according to their design specifications because the control limits are based on the mean and variability of *sampling distribution*, not the *design specifications*.

Process capability refers to the ability of the process to meet the design specifications for an output. Design specifications are often expressed as a target value (τ) and a tolerance (T). For example, the administrator of an intensive care unit lab might have a target value for the turnaround time of results to the attending physicians of 25 minutes and a tolerance of ± 5 minutes because of the need for speed under life-threatening conditions. The tolerance gives an upper specification (U) of 30 minutes and a lower specification (L) of 20 minutes. The lab process must be capable of providing the results of analyses within these specifications (see figure 4.7); otherwise, it will produce a certain proportion of "defects."

Note that in most situations,

$$T = \frac{U - L}{2} \quad \text{and} \quad \tau = \frac{U + L}{2},$$

and hence it is assumed so here.

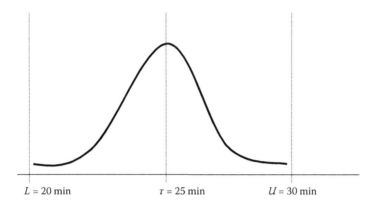

$L = 20$ min $\tau = 25$ min $U = 30$ min

FIGURE 4.7
Capable process.

Two essential quantitative measures to assess the capability of a process are process capability ratio (C_p) and process capability index (C_{pk}).

4.16.1   Process Capability Ratio (C_p)

Assume that σ is the standard deviation of a process that produces a certain dimension of interest for an output (good or service). This certain dimension of interest will hereafter be called critical dimension. The process capability ratio (C_p) is defined as

$$C_p = \frac{U - L}{6\sigma} \tag{4.69}$$

The numerator represents the specification width and the denominator captures the total width of the 3σ limits of the process distribution. We consider two examples, one for $C_p = 1$ and the other for $C_p = 2$.

If $C_p = 1$, the specification width is the same as the distribution width. When the process mean (μ) is centered at $(U+L)/2$ without any shift from the target value, τ, the probability that the actual critical dimension is within the specification limits (assuming that the process distribution is normal) is 0.9973 (2,700 ppm defect rate). Similarly, if $C_p = 2$, the specification width is twice that of the distribution. When the process mean (μ) is centered at $(U+L)/2$ without any shift from τ, the probability that the actual critical dimension is within the specification limits is 0.999999998 (0.002 ppm defect rate).

4.16.2   Process Capability Index (C_{pk})

The process capability ratio (C_p) is enough to find out whether a process is capable, only if μ is centered at $(U+L)/2$ without any shift from τ. For example, the lab process may have a good C_p value (i.e., more than the critical value of, say, 1.5), but if μ is closer to U, lengthy turnaround times may still be generated. Likewise, if μ is closer to L, very quick results may be generated. Thus, in order to check whether μ is not far away from τ, there is a need for an additional capability ratio, called the process capability index (C_{pk}).

C_{pk} is defined in [35] as

$$C_{pk} = C_p(1-k), \text{ where } k = \frac{|\tau - \mu|}{T} \tag{4.70}$$

This definition allows consideration of a mean shift, i.e., a shift of μ from τ. The fraction k is the fraction of tolerance consumed by the mean shift. The Motorola convention uses a one-sided mean shift of 1.5σ. This is motivated by common physical phenomena such as tool wear. If $C_p = 2$ and $C_{pk} = 1.5$ (i.e., mean shift consumes 25% of the tolerance), the probability that the actual critical dimension is within the specification limits is 0.9999966 (i.e., 3.4 ppm defect rate).

A process is said to be capable only if the process has good values (viz., more than the respective critical values) of both C_p and C_{pk}. If C_p is less than the critical value, σ is too high. If C_{pk} is less than the critical value, either μ is too close to U or L or σ is too high.

Six Sigma is an art of management that originated at Motorola in the early 1980s and is a business-driven, multifaceted approach to process improvement, cost reduction, and profit increase. Its fundamental principle is to improve customer satisfaction by reducing defects in processes.

Traditionally, one needs both C_p and C_{pk} values in order to investigate whether the process of interest is a Six Sigma process. We illustrate this by calculating C_p and C_{pk} values (equations (4.69) and (4.70), respectively) for an n Sigma process (where n is any positive real number; the higher the value of n, the better the process is). We consider three different cases, viz., $n = 3$, 4.5, and 6. It must be noted that the mean shift in each case is allowed to be up to 1.5σ.

4.16.2.1 Three Sigma Process

See figure 4.8.

$$C_p = \frac{U - L}{6\sigma} = \frac{6\sigma}{6\sigma} = 1$$

$$C_{pk} = C_p(1-k) \geq 1\left(1 - \frac{1.5\sigma}{3\sigma}\right) = 0.5$$

Hence, if $C_p = 1$ and $C_{pk} \geq 0.5$, it is considered a Three Sigma process.

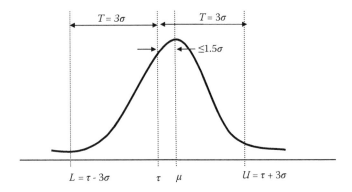

FIGURE 4.8
Three Sigma process.

4.16.2.2   4.5 Sigma Process

See figure 4.9.

$$C_p = \frac{U - L}{6\sigma} = \frac{9\sigma}{6\sigma} = 1.5$$

$$C_{pk} = C_p(1-k) \geq 1.5\left(1 - \frac{1.5\sigma}{4.5\sigma}\right) = 1$$

Hence, if $C_p = 1.5$ and $C_{pk} \geq 1$, it is considered a 4.5 Sigma process.

4.16.2.3   Six Sigma Process

See figure 4.10.

$$C_p = \frac{U - L}{6\sigma} = \frac{12\sigma}{6\sigma} = 2$$

$$C_{pk} = C_p(1-k) \geq 2\left(1 - \frac{1.5\sigma}{6\sigma}\right) = 1.5$$

Hence, if $C_p = 2$ and $C_{pk} \geq 1.5$, it is considered a Six Sigma process.

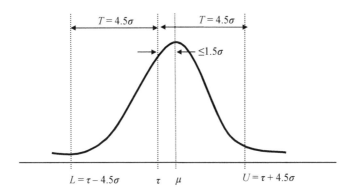

FIGURE 4.9
4.5 Sigma process.

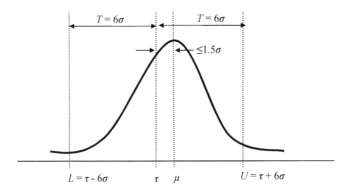

FIGURE 4.10
Six Sigma process.

4.17 Neural Networks

A neural network is made up of simple processing units (called neurons) combined in a parallel computer system following implicit instructions based on recognizing patterns in data inputs from external sources [41]. Neural networks can model complex relationships between inputs and outputs or find patterns in data. Their usefulness derives from their ability to embody inferential algorithms that alter the strengths or weights of the network connections to produce a desired significant flow. In this book, the following equation [42] is used (see chapters 6 and 7) to calculate the weights (importance values) of evaluation criteria considered in a strategic planning issue.

$$|W_v| = \dfrac{\displaystyle\sum_{j}^{n_H} \dfrac{I_{vj}}{\displaystyle\sum_{i}^{n_V} |I_{ij}|} O_j}{\left(\displaystyle\sum_{i}^{n_V} \left| \displaystyle\sum_{j}^{n_H} \left| \dfrac{I_{vj}}{\displaystyle\sum_{i}^{n_V} |I_{ij}|} O_j \right| \right| \right)}$$

(4.71)

Here, the absolute value of W_v is the weight of the vth input node (evaluation criterion) upon the output node (rating of the decision alternative), n_V is the number of input nodes (evaluation criteria), n_H is the number of hidden nodes (can be any arbitrary number), I_{ij} is the connection weight from the ith input node to the jth hidden node, and O_j is the connection weight from the jth hidden node to the output node. The connection weights [43] are obtained upon training the respective neural network.

4.18   Geographical Information Systems

A geographical information system (GIS) is a computer system with a set of processes for obtaining, managing, analyzing, and displaying data that have been located geographically [37]. In a more generic sense, GIS is a tool that allows users to create interactive queries (user-created searches), analyze the spatial information, edit data, and map and present the results of all these operations [38]. GISs have become powerful operation tools in the business world.

In this book (see chapter 6), GIS is applied for *displaying* data. The motivation for this application is a paper [39] that addresses the necessity for building the strategic planning process "around a picture," to make the chairperson of the concerned supply chain company easily understand the "dense documents filled with numbers" and to convince him or her that it is important to implement the proposed action. To this end, an excellent GIS-based business mapping application, MapLand [40], is used to map the results obtained in different phases of a model.

4.19   Linear Integer Programming

Linear programming problems are about optimization of a linear objective function, subject to linear equality and inequality constraints [36]. For example, in canonical form, a linear programming problem can be expressed as

Maximize $C^T X$ (objective function)
subject to $AX \leq B$ and $X \geq 0$ (constraints)

Here, X represents the vector of variables, whereas C and B are vectors of coefficients and A is the matrix of coefficients. If the variables of a linear programming problem are restricted to being integers, the programming is called linear integer programming.

Linear integer programming can be applied to various fields. Most extensively, it is applied to business, economic, and engineering problems.

4.20   Conclusions

This chapter gave an introduction to various quantitative techniques employed by strategic planning models presented in different chapters of the book. Depending on the decision-making situation, each of the models uses one or more of the techniques introduced in this chapter.

References

1. Saaty, T. L. 1980. *The analytic hierarchy process*. New York: McGraw-Hill.
2. Siddiqui, M. Z., Everett, J. W., and Vieux, B. E. 1996. Landfill siting using geographic information systems: A demonstration. *Journal of Environmental Engineering* 122:515–23.
3. Saaty, T. L. 1990. How to make a decision: The analytic hierarchy process. *European Journal of Operational Research* 48:9–26.
4. Tadisina, S. K., Troutt, M. D., and Bhasin, V. 1991. Selecting a doctoral programme using the analytic hierarchy process: The importance of perspective. *Journal of the Operational Research Society* 42:631–38.
5. Prakash, G. P., Ganesh, L. S., and Rajendran, C. 1994. Criticality analysis of spare parts using the analytic hierarchy process. *International Journal of Production Economics* 35:293–97.
6. Saaty, T. L. 1996. *Decision making with dependence and feedback: The analytic network process*. Pittsburgh, PA: RWS Publications.

7. Ravi, V., Shankar, R., and Tiwari, M. K. 2005. Analyzing alternatives in reverse logistics for end-of-life computers: ANP and balanced scorecard approach. *Computers and Industrial Engineering* 48:327–56.
8. Gungor, A. 2006. Evaluation of connection types in design for disassembly (DFD) using analytic network process. *Computers and Industrial Engineering* 50:35–54.
9. Agarwal, A., Shankar, R., and Tiwari, M. K. 2006. Modeling the metrics of lean, agile and leagile supply chain: An ANP based approach. *European Journal of Operational Research* 173:211–25.
10. Zadeh, L. A. 1965. Fuzzy sets. *Information and Control* 8:338–53.
11. Tsaur, S., Chang, T., and Yen, C. 2002. The evaluation of airline service quality by fuzzy MCDM. *Tourism Management* 23:107–15.
12. Chan, F. T., Chan, H. K., and Chan, M. H. 2003. An integrated fuzzy decision support system for multi-criterion decision making problems. *Journal of Engineering Manufacture* 217:11–27.
13. Wang, M., and Liang, G. 1995. Benefit/cost analysis using fuzzy concept. *Engineering Economist* 40:359–76.
14. Chang, D. Y. 1996. Applications of the extent analysis method on fuzzy AHP. *European Journal of Operations Research* 95:649–55.
15. Chang D. Y. 1992. *Extent analysis and synthetic decision, optimization techniques and applications*, p. 352. Vol. 1. Singapore: World Scientific.
16. Chen, S. J., and Hwang, C. L. 1992. *Fuzzy multiple attribute decision making: Methods and applications*. New York: Springer.
17. Deng, H. 1999. Multicriteria analysis with fuzzy pairwise comparison. *International Journal of Approximate Reasoning* 21:215–31.
18. Zeleny, M. 1982. *Multiple criteria decision making*. New York: McGraw-Hill.
19. Hwang, C. L., and Yoon, K. S. 1981. *Multiple attribute decision making: Methods and applications*. Berlin: Springer.
20. Yeh, C. H., and Deng, H. 1997. An algorithm for fuzzy multi-criteria decision making. In *Proceedings of the IEEE First International Conference on Intelligent Processing Systems*, pp. 1564–68.
21. Erol, I., and Ferrell, Jr., W. G. 2003. A methodology for selecting problems with multiple, conflicting objectives and both qualitative and quantitative data. *International Journal of Production Economics* 86:187–99.
22. Messac, A., Gupta, S. M., and Akbulut, B. 1996. Linear physical programming: A new approach to multiple objective optimization. *Transactions on Operational Research* 8:39–59.
23. Martinez, M., Messac, A., and Rais-Rohani, M. 2001. Manufacturability-based optimization of aircraft structures using physical programming. *AIAA Journal* 39:517–25.
24. Hillier, F. S., and Lieberman, G. J. 2001. *Introduction to operations research*. New York: McGraw-Hill.
25. Ignizio, J. P. 1976. *Goal programming and extensions*. Lexington, MA: Lexington Books, D. C. Heath and Company.
26. Ignizio, J. P. 1982. *Linear programming in single and multi objective systems*. Englewood Cliffs, NJ: Prentice Hall.
27. Tiwari, R. N., Dharmar, S., and Rao, J. R. 1987. Fuzzy goal programming—An additive model. *Fuzzy Sets and Systems* 24:27–34.
28. Zimmermann, H. J. 1978. Fuzzy programming and linear programming with several objective functions. *Fuzzy Sets and Systems* 1:45–55.

29. Triantaphyllou, E., and Lin, C. 1996. Development and evaluation of five fuzzy multi-attribute decision-making methods. *International Journal of Approximate Reasoning* 14:281–310.
30. Hwang, C. L. 1987. *Group decision making under multi-criteria: Methods and applications.* New York: Springer-Verlag.
31. Rich, E., and Knight, K. 1992. *Artificial intelligence.* New York: McGraw-Hill.
32. Hopgood, A. A. 1993. *Knowledge-based systems for engineers and scientists.* Boca Raton, FL: CRC Press.
33. Pi, W., and Low, C. 2006. Supplier evaluation and selection via Taguchi loss functions and an AHP. *International Journal of Advanced Manufacturing Technology* 27:625–30.
34. Krajewski, L. J., and Ritzman, L. P. 2004. *Operations management: Processes and value chains.* 7th ed. Englewood Cliffs, NJ: Pearson Prentice Hall.
35. Narahari, Y., Viswanadham, N., and Bhattacharya, R. 2000. Design of synchronized supply chains: A Six Sigma tolerancing approach. In *IEEE International Conference on Robotics and Automation,* pp. 1151–56.
36. Hillier, F. S., and Lieberman, G. J. 2005. *Introduction to operations research.* 8th ed. New York: McGraw-Hill.
37. Estaville, L. E. 2007. GIS and colleges of business: A curricular exploration. *Journal of Real Estate Literature* 15:443–48.
38. http://en.wikipedia.org/wiki/Geographic_information_system.
39. Kim, C. W., and Mauborgne, R. 2002. Charting your company's future. *Harvard Business Review,* June, pp. 77–82.
40. http://www.softwareillustrated.com/mapland.html.
41. Masi, C. G. 2007. Fuzzy neural control systems—explained. *Control Engineering* 54:62–66.
42. Cha, Y., and Jung, M. 2003. Satisfaction assessment of multi-objective schedules using neural fuzzy methodology. *International Journal of Production Research* 41:1831–49.
43. Haykin, S. 1998. *Neural networks: A comprehensive foundation.* 2nd ed. Englewood Cliffs, NJ: Prentice Hall.

5

Selection of Used Products

5.1 The Issue

In many countries, especially in Europe, although many original equipment manufacturers (OEMs) are obligated to take products back from the consumers upon the products' end of use (hence called *used products*), there are also many third-party companies that collect used products solely to make profit. These companies select only those used products for which revenues from recycle or resale of the products' components are expected to be higher than the costs involved in collection and reprocessing of used products and in disposal of waste. The various scenarios for selecting economical used products could differ as follows:

1. Evaluation criteria could be presented in terms of classical numerical constraints.
2. Evaluation criteria could be presented in terms of ranges of different degrees of desirability.

In this chapter, two models to address the above scenarios are presented. The first model addresses scenario 1 and employs linear integer programming. The second model is for scenario 2 and employs linear physical programming.

This chapter is organized as follows: section 5.2 presents the first model, section 5.3 shows the second model, and section 5.4 gives some conclusions.

5.2 First Model (Linear Integer Programming)

Section 5.2.1 presents the nomenclature for formulation of the linear integer programming model, section 5.2.2 presents the formulation of the model, and section 5.2.3 gives a numerical example to illustrate the model.

5.2.1   Nomenclature

C_{df}	Disposal cost factor (cost per unit weight)
C_{dx}	Disposal cost of product x
C_r	Reprocessing cost per unit time
C_{rf}	Recycling revenue factor (revenue per unit weight)
C_{rpx}	Total reprocessing cost of used product x
CC_x	Collection cost of used product x
D_{xy}	Disposal cost index of component y in used product x (0 = lowest, 10 = highest)
E_{xk}	Subassembly k in used product x
m_{xy}	Probability of missing component y in used product x
M_x	Number of subassemblies in used product x
N_{xy}	Multiplicity of component y in used product x
p_{xy}	Probability of breakage of component y in used product x
P_{xy}	Component y in used product x
PRC_{xy}	Percent of recyclable contents by weight in component y of used product x
R_{rcx}	Total recycling revenue of used product x
R_{rsx}	Total resale revenue of used product x
R_{rsxy}	Resale value of component y in used product x
RC_{xy}	Recycling revenue index of component y in used product x (0 = lowest, 10 = highest)
$Root_x$	Root node of used product x
SU_x	Supply of used product x per period
$T(E_{xk})$	Time to disassemble subassembly k in used product x
$T(Root_x)$	Time to disassemble $Root_x$
W_{xy}	Weight of component y in used product x
x	Used product type
X_{xy}	Decision variable signifying selection of component y to be retrieved from used product x for reuse ($X_{xy}=1$ for reuse, 0 for recycle)
y	Component type
Z	Overall profit

5.2.2   Model Formulation

The cost-benefit function [1] in the literature presents a technique to select the best used product for reprocessing from a set of candidate used products. The function consists of four terms: resale revenue, recycling revenue, reprocessing cost, and disposal cost. The difference between the sum of the revenue terms and the sum of the cost terms gives the cost-benefit function value. The used product with the maximum cost-benefit function value is selected as the optimal one for reprocessing. The optimal solution (for profit maximization) also gives the best feasible set of components of retrieval from the selected used product. However, there is serious risk of making a bad choice of the used product using the cost-benefit function because the function implicitly makes two assumptions that are seldom valid in a reverse supply chain scenario: (1) every component selected for resale will be in a reusable state after disassembling the used product, and (2) that all the components in the used product are in their original multiplicities. Therefore, here, a modified cost-benefit function is presented. This function incorporates the probability of breakage and the probability of missing components in the used product of interest. Then a linear integer programming model is formulated and implemented to the data of each candidate used product, in order to select the most economical used product to reprocess. Section 5.2.2.1 presents the modified cost-benefit function, and section 5.2.2.2 gives the linear integer programming model.

5.2.2.1   Modified Cost-Benefit Function

The modified cost-benefit function for used product x consists of five terms: total resale revenue (R_{rsx}), total recycling revenue (R_{rcx}), total reprocessing cost (C_{rpx}), total disposal cost (C_{dx}), and collection cost. It can be written as

$$Z = R_{rsx} + R_{rcx} - C_{rpx} - C_{dx} - CC_x \qquad (5.1)$$

The terms in the function are described as follows:

Total resale revenue: R_{rsx} is influenced by the resale value of individual components of the used product (R_{rsxy}), the number of components (N_{xy}), the probability of breakage (p_{xy}), and the probability of missing components (m_{xy}). The revenue equation can be written as follows:

$$R_{rsx} = \sum_{j \ni Pxy \in (Root_x)} \left\{ R_{rsxy} . N_{xy} . (1 - p_{xy} - m_{xy}) . X_{xy} \right\} \qquad (5.2)$$

Total recycling revenue: R_{rcx} is influenced by the percentage of recyclable contents in each component (PRC_{xy}); the weight of the components (W_{xy}); the recycling revenue index (RC_{xy}), which is a number in the range 1–10 representing the degree of benefit generated by recycling a component of type y (the higher the value, the more profitable it is to recycle); the number of components (N_{xy}); the probability of breakage (p_{xy}); and the probability of missing components (m_{xy}). The recycling revenue equation can be written as follows:

$$R_{rcx} = \sum_{j \ni Pxy \in (Root_x) \;\}} \left\{ PRC_{xy}.W_{xy}.RC_{xy}(N_{xy}(1-m_{xy})-N_{xy}(1-p_{xy}-m_{xy})X_{xy}) \right\} \times C_{rf}$$

(5.3)

Total reprocessing cost: C_{rpx} can be calculated from the disassembly time of the root node of the used product ($T(Root_x)$), the disassembly time of each subassembly in the used product ($T(E_{xk})$), and the reprocessing cost per unit time (C_r). The total reprocessing cost equation can be written as follows:

$$C_{rpx} = \left\{ T(Root_x) + \sum_{k=1}^{M_x} T(E_{xk}) \right\}.C_r$$

(5.4)

Total disposal cost: C_{dx} can be calculated from the disposal cost index (D_{xy}), which is a number in the range 1–10 representing the degree of difficulty in disposing component y of used product x (the higher the number, the more difficult it is to dispose), the percentage of recyclable contents in each component (PRC_{xy}), the weight of the components (W_{xy}), the number of components (N_{xy}), the probability of breakage (p_{xy}), the probability of missing components (m_{xy}), and the disposal cost factor (C_{df}). The disposal cost equation can be written as follows:

$$C_{dx} = \sum_{j \ni Pxy \in (Root_x)} \left\{ D_{xy}.W_{xy}.(1-PRC_{xy}).(N_{xy}(1-m_{xy})-N_{xy}(1-p_{xy}-m_{xy}).X_{xy}) \right\}.C_{df}$$

(5.5)

Collection cost: CC_x is the average cost of collecting used product x from the consumers.

5.2.2.2 Linear Integer Programming Model

The following linear integer programming model maximizes the cost-benefit obtained from reprocessing used product x:

$$\text{Maximize } Z_x = R_{rsx} + R_{rcx} - C_{rpx} - C_{dx} - CC_x$$

(5.6)

$$\text{subject to } X_{xy} = 0 \text{ or } 1 \text{ for all } x \text{ and } y$$

(5.7)

The above formulation assesses the feasible combinations of components' retrieval from a used product and compares the combination with the highest cost-benefit from one product against others.

5.2.3 Numerical Example

Two used products (1 and 2) are considered in this numerical example (see figures 5.1 and 5.2).

The data required to implement the model for the products are shown in tables 5.1 and 5.2, respectively.

Also, $CC_1 = 25$, $CC_2 = 40$, $C_{rf} = 0.5$, $C_r = 0.8/min$, $C_{df} = 0.25/lb$, $T(Root_1) = 6$ min, $T(Root_2) = 4$ min, $T(E_{11}) = 3$ min, $T(E_{12}) = 5$ min, $T(E_{21}) = 2$ min, and $T(E_{22}) = 4$ min.

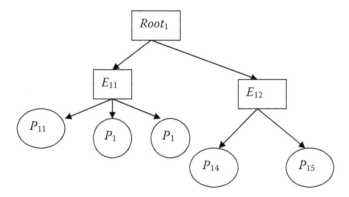

FIGURE 5.1
Structure of used product 1 (first model).

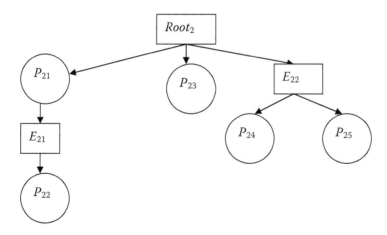

FIGURE 5.2
Structure of used product 2 (first model).

TABLE 5.1

Data of used product 1 (first model)

Part	R_{rs1y}	N_{1y}	W_{1y}	$Rc1y$	PRC_{1y}	D_{1y}	p_{1y}	m_{1y}
P_{11}	10	4	3	7	75%	3	0.02	0.0
P_{12}	2	2	2.5	9	60%	2	0.05	0.0
P_{13}	3.75	1	5	5	50%	6	0.0	0.5
P_{14}	5	5	7	3	70%	1	0.1	0.3
P_{15}	3	6	4	6	40%	7	0.4	0.02

TABLE 5.2

Data of used product 2 (first model)

Part	R_{rs2y}	N_{2y}	W_{2y}	R_{c2y}	PRC_{2y}	D_{2y}	p_{2y}	m_{2y}
P_{21}	2.5	2	4	5	40%	2	0.0	0.05
P_{22}	5	3	1	6	35%	6	0.0	0.1
P_{23}	3	1	3	2	70%	8	0.1	0.0
P_{24}	0.5	2	4.5	8	80%	5	0.15	0.2
P_{25}	2	2	5	7	25%	7	0.1	0.25

Upon solving the model with the above data using LINGO (v4), one gets total profit for used product 1 as \$46.16 and total profit for used product 2 as \$68.50. Hence, the decision maker will select used product 2 in this case.

5.3   Second Model (Linear Physical Programming)

Section 5.3.1 presents the formulation of the linear physical programming model, and section 5.3.2 gives a numerical example to illustrate the model. The nomenclature used to formulate the model is the same as in section 5.2.1.

5.3.1   Model Formulation

The criteria considered in the model fall into either class 1S or class 2S. Section 5.3.1.1 presents the class 1S criteria, and section 5.3.1.2 gives the class 2S criteria.

5.3.1.1   *Class 1S Criteria (Smaller Is Better)*

Total collection cost per period (g_1): g_1 of used product x is calculated by multiplying the supply of x per period (SU_x) by the cost of collecting one product from consumers (CC_x):

$$g_1 = SU_x \times CC_x \tag{5.8}$$

Total reprocessing cost per period (g_2): g_2 of used product x is calculated using the disassembly time of the root node ($T(Root_x)$), disassembly time of each subassembly ($T(E_{xk})$), supply of x per period (SU_x), and the remanufacturing cost per unit time (C_r):

$$SU_x \left[T(Root_x) + \sum_{k=1}^{M_i} T(E_{xk}) \right] C_r \tag{5.9}$$

Total disposal cost per period (g_3): g_3 of used product x is calculated by multiplying the component disposal cost by the number of units of components disposed, as follows:

$$\sum_y \left[SU_x.DI_{xy}W_{xy}(1 - PRC_{xy}) \{ N_{xy}(1 - m_{xy}) - N_{xy}(1 - b_{xy} - m_{xy}) \} \right] C_{df} \tag{5.10}$$

DI_{xy} is the disposal cost index that varies in value from 1 to 10 representing the degree of nuisance created by the disposal of component y of product x, and C_{df} is the disposal cost factor.

Loss-of-sale cost (g_4): g_4 of used product x represents the periodic worth of not meeting the demand on time. It can occur because of the unpredictability in the supply of used products. This can be obtained from an expert in the field.

Worth of investment cost (g_5): g_5 of used product x represents the periodic worth of the fixed cost of the production facility and the machinery required to reprocess. This can also be obtained from an expert in the field.

5.3.1.2 Class 2S Criteria (Larger Is Better)

Total reuse revenue per period (g_6): g_6 of used product x is influenced by the supply of x per period (SU_x), the resale value of component y (RSR_{xy}), the multiplicity of component y (N_{xy}), and the breakage and missing probabilities of component y (p_{xy}, m_{xy}). The reuse revenue equation can be written as follows:

$$\sum_y \left[SU_x.RSRxy.N_{xy}.(1 - m_{xy} - b_{xy}) \right]$$

$$\tag{5.11}$$

Total recycling revenue per period (g_7): g_7 of used product x is influenced by the supply of x per period (SU_x), the recycling revenue index of component y ($RCRI_{xy}$), the percentage of recyclable content in component y (PRC_{xy}), the multiplicity of component y (N_{xy}), the weight of component y (W_{xy}), the recycling revenue factor (C_{rf}), and the breakage and missing probabilities of component y (p_{xy}, m_{xy}). The recycling revenue equation can be written as

$$\sum_{y}\left[SU_x . RCRI_{xy} . W_{xy} . PRC_{xy} . N_{xy} . \left\{N_{xy}(1-m_{xy}) - N_{xy}(1-m_{xy}-b_{xy})\right\}C_{rf}\right] \quad (5.12)$$

5.3.2 Numerical Example

Consider three used products (3, 4, and 5), whose structures are shown in figures 5.3–5.5, respectively.

The data for the three products are given in tables 5.3–5.5, respectively. The target values for the criteria are given in table 5.6, and table 5.7 shows the criteria values for each product. Table 5.8 shows the incremental weights obtained using the LPP weight algorithm [2].

Tables 5.9–5.11 show the deviations of criteria values from target values for the three products, respectively. For example, for criteria g_1 of product 3, deviation at $s = 2$ is the absolute value of the number obtained by subtracting the criteria value (i.e., 7.5; shown in bold in table 5.7) from the target value (i.e., 10; shown in bold in table 5.6).

The total score for each product is calculated using equation (4.34) (using the incremental weights from the LPP algorithm and deviations from target values) and is shown in table 5.12. Because alternatives with lower scores are more desirable than ones with higher scores, used product 3 is the best of the lot.

5.4 Conclusions

In this chapter, two models are presented for selecting the most economical product to reprocess from a set of candidate used products. The first model uses linear integer programming, and the second model employs linear physical programming.

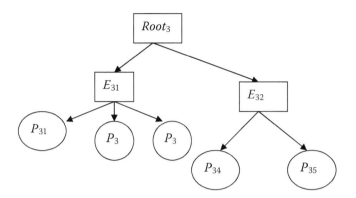

FIGURE 5.3
Structure of used product 3 (second model).

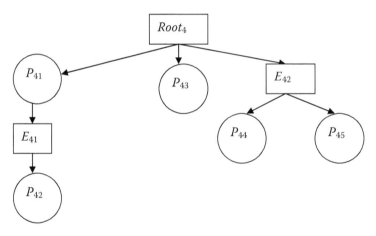

FIGURE 5.4
Structure of used product 4 (second model).

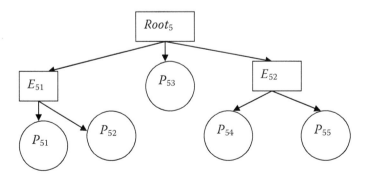

FIGURE 5.5
Structure of used product 5 (second model).

TABLE 5.3

Data of used product 3 (second model)

Part	RSR_{xy}	N_{xy}	W_{1y}	$RCRI_{xy}$	PRC_{xy}	DI_{xy}	p_{xy}	m_{xy}
P_{31}	6	3	3.5	5	0.65	6	0.1	0.3
P_{32}	7	2	4.5	4	0.4	4	0.2	0.5
P_{33}	5	4	5	5	0.3	4	0.4	0.1
P_{34}	7.5	2	6	3	0.25	5	0.2	0.2
P_{35}	5.5	1	5.5	2	0.45	1	0.5	0.0

TABLE 5.4

Data of used product 4 (second model)

Part	RSR_{xy}	N_{xy}	$W2_{y}$	$RCRI_{xy}$	PRC_{xy}	DI_{xy}	p_{xy}	m_{xy}
P_{41}	4	2	9	2	0.56	3	0.2	0.5
P_{42}	6	5	3	1	0.5	6	0.2	0.2
P_{43}	7	4	5	2	0.48	6	0.4	0.1
P_{44}	2	2	6	3	0.2	1	0.1	0.1
P_{45}	4.5	3	1	4	0.25	3	0.1	0.3

TABLE 5.5

Data of used product 5 (second model)

Part	RSR_{xy}	N_{xy}	$W3_{y}$	$RCRI_{xy}$	PRC_{xy}	DI_{xy}	p_{xy}	m_{xy}
P_{51}	4.5	3	8.5	5	0.4	5	0.1	0.3
P_{52}	4	4	6	4	0.65	6	0.1	0.1
P_{53}	5	6	4.1	2	0.2	3	0.2	0.0
P_{54}	2	5	3.5	3	0.35	2	0.3	0.2
P_{55}	3.5	1	2	1	0.25	7	0.5	0.5

TABLE 5.6

Target values of criteria (second model)

Criteria	$t_{p1}+$	$t_{p2}+$	$t_{p3}+$	$t_{p4}+$	$t_{p5}+$
g_1	10	12	15	17.5	20
g_2	3	5	9	10	15
g_3	1	4	5.5	7.5	8
g_4	2	5	7.5	9	10
g_5	1	4	8	9	10
Criteria	$t_{p1}-$	$t_{p2}-$	$t_{p3}-$	$t_{p4}-$	$t_{p5}-$
g_6	10	12.5	15	20	25.5
g_7	5	13	17.5	20	25

TABLE 5.7

Criteria values for each product (second model)

Criteria	Used product 3	Used product 4	Used product 5
g_1	7.5	7.5	10
g_2	3.5	2.75	3
g_3	2.97	1.6	1.8
g_4	3.5	4	5
g_5	2	2.5	2
g_6	18.3	22.8	24.9
g_7	18.4	15.7	19

TABLE 5.8

Output of LPP weight algorithm (second model)

Criteria	$\Delta w_{p2}+$	$\Delta w_{p3}+$	$\Delta w_{p4}+$	$\Delta w_{p5}+$	$\Delta w_{p2}-$	$\Delta w_{p3}-$	$\Delta w_{p4}-$	$\Delta w_{p5}-$
g_1	0.05	0.0967	0.62773	2.63296	—	—	—	—
g_2	0.05	0.076	2.41416	0.020321	—	—	—	—
g_3	0.0333	0.3027	0.93408	24.33473	—	—	—	—
g_4	0.0333	0.16827	1.49184	11.10897	—	—	—	—
g_5	0.0333	0.09267	2.41416	10.26225	—	—	—	—
g_6	—	—	—	—	0.04	0.0492	0.010258	0.122333
g_7	—	—	—	—	0.0125	0.037056	0.14936	0.022875

TABLE 5.9

Deviations of criteria values from targets, for used product 3 (second model)

Criteria	$s = 2$	$s = 3$	$s = 4$	$s = 5$
g_1	**2.5**	4.5	7.5	10
g_2	0.5	1.5	5.5	6.5
g_3	1.9	1.1	2.6	4.6
g_4	1.5	1.5	4	5.5
g_5	1	2	6	7
g_6	6.7	1.7	3.3	5.8
g_7	6.6	1.6	0.9	5.4

TABLE 5.10

Deviations of criteria values from targets, for used product 4 (second model)

Criteria	$s = 2$	$s = 3$	$s = 4$	$s = 5$
g_1	2.5	4.5	7.5	10
g_2	0.25	2.25	6.25	7.25
g_3	0.6	2.4	3.9	5.9
g_4	2	1	3.5	5
g_5	1.5	1.5	5.5	6.5
g_6	2.2	2.8	7.8	10.3
g_7	9.3	4.3	1.8	2.7

TABLE 5.11

Deviations of criteria values from targets, for used product 5 (second model)

Criteria	$s = 2$	$s = 3$	$s = 4$	$s = 5$
g_1	0	2	5	7.5
g_2	0	2	6	7
g_3	0.8	2.2	3.7	5.7
g_4	3	0	2.5	4
g_5	1	2	6	7
g_6	0.1	4.9	9.9	12.4
g_7	5.93	0.93	1.57	6.07

TABLE 5.12
Total scores and ranks of products (second model)

Used product	Score	Rank
3	315.3144	1
4	339.1954	3
5	317.8653	2

References

1. Veerakamolmal, P., and Gupta, S. M. 1999. Analysis of design efficiency for the disassembly of modular electronic products. *Journal of Electronics Manufacturing* 9:79–95.
2. Messac, A., Gupta, S. M., and Akbulut, B. 1996. Linear physical programming: A new approach to multiple objective optimization. *Transactions of Operational Research* 8:39–59.

6

Evaluation of Collection Centers

6.1 The Issue

Strategic planning of an efficient reverse or closed-loop supply chain requires selection of efficient collection centers where used products are disposed of by the consumers. These collection centers, after initial processing (for example, sorting), ship the used products to recovery facilities or production facilities where reprocessing operations, such as disassembly and recycling/remanufacturing, are carried out.

The various scenarios for evaluating collection centers for efficiency could differ as follows:

1. Supply chain company executives, whose primary concern is profit, could be the sole decision makers.

2. There could exist three different categories of decision makers: consumers, local government officials, and supply chain company executives. Impacts* of evaluation criteria are given.

3. The situation could be the same as in scenario 2, but the impacts of evaluation criteria are not given (and hence must be derived).

4. Evaluation could be made from the perspective of a remanufacturing facility interested in buying used products from the candidate collection centers. The goals are expressed in terms of performance indices (efficiency scores).

5. The situation could be the same as in scenario 4, but the goals are expressed in terms of Taguchi losses (inefficiency scores).

In this chapter, various models to address the above scenarios are presented. The first model addresses scenario 1 and employs the eigen vector method and Taguchi loss function. The second and third models are for scenarios 2 and 3, respectively. The main difference between these two models

---

* In this chapter, we use the term *impacts* instead of the conventional term *weights* for importance values of evaluation criteria. This is to avoid confusion in the presence of the connection weights between different nodes of the neural network used in section 6.5.

lies in the way the impacts are assigned to the criteria for evaluation of collection centers. Whereas the second model uses the eigen vector method, technique for order preference by similarity to ideal solution (TOPSIS), and Borda's choice rule, the third model uses neural networks, fuzzy logic, TOPSIS, and Borda's choice rule. The fourth model is for scenario 4 and uses analytic network process (ANP) and goal programming. The fifth model, which is for scenario 5, uses the eigen vector method, Taguchi loss function, and goal programming. The main difference between the last two models is in the way the goals are expressed.

This chapter is organized as follows: Section 6.2 presents the first model. Section 6.3 gives the different decision makers (and their criteria) that are considered in the second and third models. Sections 6.4–6.7 present the second, third, fourth, and fifth models, respectively. Finally, section 6.8 gives some conclusions.

6.2   First Model (Eigen Vector Method and Taguchi Loss Function)

In this model, the eigen vector method and Taguchi loss function are employed to identify efficient centers from a set of candidate collection centers operating in a region where a reverse supply chain is to be designed. Section 6.2.1 presents the evaluation criteria used in the model, and section 6.2.2 illustrates the model using a numerical example.

6.2.1   Evaluation Criteria

This model assumes that the supply chain company executives, whose primary concern is profit, are the sole decision makers. Hence, the evaluation criteria do not consider factors such as "Is the collection center able to provide employment opportunities to the local community?" and "Is the collection process convenient to the consumers?" (For factors such as these, see sections 6.4 and 6.5, where criteria of two additional different decision makers, viz., consumers and local government officials, are considered.)

The following are the evaluation criteria used in the model:

- n value (the higher the n value, the lower the process defects, and hence the higher the profit; see section 4.16 for the concept of n Sigma). The n value of a collection center represents the center's quality that could be a function of factors, such as functionality of used products collected, efficiency of collection, efficiency of delivery (to recovery facilities), and effectiveness of customer service. For example (see figure 6.1), let the upper specification (U) of the service time (critical

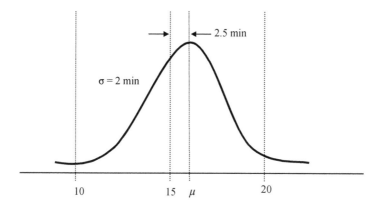

FIGURE 6.1
Critical dimension of service process at candidate collection center (first model).

dimension) at a candidate collection center be 20 min (i.e., the consumers discarding used products must not be made to wait too long at the collection center), the lower specification (*L*) be 10 units (i.e., the collection center personnel must spend at least some time inspecting used products before accepting them from consumers), and the target value (τ) be 15 min. If the standard deviation (σ) of the service time is 2 min and the mean shift ($|\tau - \mu|$) is 2.5 min, then the service process in the candidate collection center is a 2.5 Sigma process.

- Per capita income of people in residential area (*PI*) (the higher it is, the greater the number of resourceful discarded products and the less the people will care about the incentives from the collection center)
- Space cost (*SC*) (the lower, the better)
- Labor cost (*LC*) (the lower, the better)
- Utilization of incentives from local government (*UI*) (the higher, the better)
- Distance from residential area (*DH*) (lower distance implies greater collection and hence greater profit)
- Distance from roads (*DR*) (lower distance implies greater collection and hence greater profit)
- Incentives from local government (*IG*) (higher incentives from local government imply higher incentives to consumers, and hence greater collection)

6.2.2 Model

The model is presented using a numerical example. Table 6.1 shows the pairwise comparison matrix for the criteria for evaluation of candidate collection

TABLE 6.1

Pair-wise comparison matrix (first model)

Criteria	n	DH	DR	UI	PI	SC	LC	IG
n	1	1	1	3	1/5	1	1	1
DH	1	1	1	1	1	1	1	1
DR	1	1	1	1	1	1	1	1
UI	1/3	1	1	1	1/3	0.2	1	1
PI	5	1	1	3	1	1	1	1
SC	1	1	1	5	1	1	1	0.2
LC	1	1	1	1	1	1	1	1
IG	1	1	1	1	1	5	1	1

centers. For example (see table 6.1), the per capita income of the people in the residential area (*PI*) is given five times more importance than the n value and three times more importance than the utilization of incentives from the local government (*UI*).

Table 6.2 shows the impacts of the respective evaluation criteria. These impacts are the elements of the normalized eigen vector of the pair-wise comparison matrix shown in table 6.1. For example (see table 6.2), the per capita income of the people in the residential area (*PI*) is given an impact of 18% and the n value is given an impact of 11%.

The k value of the Taguchi loss function, in the case of each of the evaluation criteria, is calculated as follows.

6.2.2.1   n Value

The loss function that applies to this criterion is "larger is better" (see equation (4.68)). If the decision maker considers 100% loss for an n value less than or equal to 4, then the value of k is 1,600%. Then, for a candidate collection center, if $n = 5$, $L(y) = 1600 / (5)^2 = 64\%$. That means, with respect to quality

TABLE 6.2

Impacts of criteria (first model)

Criteria	Impacts
n	0.11
DH	0.11
DR	0.11
UI	0.07
PI	0.18
SC	0.15
LC	0.11
IG	0.16

(i.e., n value), the collection center is 36% short of the worst performance level (which, in this case, is $n \leq 4$).

6.2.2.2 Distance from Residential Area (DH)

DH is considered the distance of the collection center from the center of gravity [1] of all the residential areas around the center. The loss function that applies to this criterion is "smaller is better" (see equation (4.67)). If the decision maker considers 100% loss for a DH value more than or equal to 4 miles, then the value of k is 6.25%. Then, for a candidate collection center, if $DH = 3$, $L(y) = 6.25 \times (3)^2 = 56.25\%$. That means, with respect to the distance from the residential area, the collection center is 43.75% short of the worst-case scenario (which, in this case, is $DH \geq 4$ miles).

6.2.2.3 Distance from Roads (DR)

DR is considered the average distance of all the roads in the region from the collection center of interest. The loss function that applies to this criterion is "smaller is better" (see equation (4.67)). If the decision maker considers 100% loss for a DR value more than or equal to 5 miles, then the value of k is 4%. Then, for a candidate collection center, if $DR = 4$, $L(y) = 4 \times (4)^2 = 64\%$. That means, with respect to the distance from the roads, the collection center is 36% short of the worst-case scenario (which, in this case, is $DR \geq 5$ miles).

6.2.2.4 Utilization of Incentives from Local Government (UI)

Because this is a subjective criterion, the decision maker can obtain ratings (for example, on a 1–10 scale, where 1 is the worst and 10 is the best) of the candidate collection centers from experts in the field of reverse supply chain. The loss function that applies to this criterion is "larger is better" (see equation (4.68)). If the decision maker considers 100% loss for a UI value less than or equal to 5, then the value of k is 2,500%. Then, for a candidate collection center, if $UI = 7$, $L(y) = 2500/(7)^2 = 51.02\%$. That means, with respect to the utilization of incentives from the local government, the collection center is about 49% short of the worst-case scenario (which, in this case, is $UI \leq 5$).

6.2.2.5 Per Capita Income of People in Residential Area (PI)

The loss function that applies to this criterion is "larger is better" (see equation (4.68)). If the decision maker considers 100% loss for a PI value less than or equal to $30,000 per year, then the value of k is 90,000,000,000%. Then, for a candidate collection center, if $PI = $50,000 per year, $L(y) = 90,000,000,000/(50,000)^2 = 36\%$. That means, with respect to the per capita income of the people in the residential area, the collection center is 64% short of the worst-case scenario (which, in this case, is $PI \leq $30,000 per year).

6.2.2.6 Space Cost (SC)

The loss function that applies to this criterion is "smaller is better" (see equation (4.67)). If the decision maker considers 100% loss for a SC value more than or equal to $1,000 per day, then the value of k is 0.0001%. Then, for a candidate collection center, if SC = $800 per day, $L(y)=0.0001\times(800)^2=64\%$. That means, with respect to the space cost, the collection center is 36% short of the worst-case scenario (which, in this case, is $SC \geq$ $1,000 per day).

6.2.2.7 Labor Cost (LC)

The loss function that applies to this criterion is "smaller is better" (see equation (4.67)). If the decision maker considers 100% loss for a LC value more than or equal to $15 per hour, then the value of k is 0.44%. Then, for a candidate collection center, if LC = $10 per hour, $L(y)=0.44\times(10)^2=44\%$. That means, with respect to the labor cost, the collection center is 56% short of the worst-case scenario (which, in this case, is $LC \geq$ $15 per hour).

6.2.2.8 Incentives from Local Government (IG)

Because this is a subjective criterion, the decision maker can obtain ratings (for example, on a 1–10 scale, where 1 is the worst and 10 is the best) of the candidate collection centers from experts in the field of reverse supply chain. The loss function that applies to this criterion is "larger is better" (see equation (4.68)). If the decision maker considers 100% loss for an IG value less than or equal to 7, then the value of k is 4,900%. Then, for a candidate collection center, if IG = 9, $L(y)=4900/(9)^2=60.49\%$. That means, with respect to the incentives from the local government, the collection center is about 39.5% short of the worst-case scenario (which, in this case, is $IG \leq 7$).

Four candidate collection centers are considered in the numerical example: A, B, C, and D. Table 6.3 presents the $L(y)$ for the evaluation criteria for each of the collection centers. For example, the Taguchi loss of B with respect to the per capita income of the people in the residential area is 52% (i.e., 48% short of the worst-case scenario).

The weighted loss of each collection center j is calculated by using the following equation and is presented in table 6.4.

$$\text{Weighted loss of collection center } j = \sum_i W_i L_{ij} \qquad (6.1)$$

where W_i is the impact of criterion i (see table 6.2) and L_{ij} is the Taguchi loss (see table 6.3) of collection center j with respect to criterion i. For example, the weighted loss of D (see tables 6.2–6.4) is 0.11×63 + 0.11×55 + … + 0.11×67 + 0.16×41 = 64.31.

The decision maker will select C because it has the lowest weighted loss.

TABLE 6.3

$L(y)$ values (%) of collection centers (first model)

Criteria	A	B	C	D
n (0.11)	35	54	40	63
DH (0.11)	25	15	37	55
DR (0.11)	32	10	9	90
UI (0.07)	100	75	64	50
PI (0.18)	65	52	40	50
SC (0.15)	20	10	5	100
LC (0.11)	15	18	20	67
IG (0.16)	78	64	36	41

TABLE 6.4

Weighted losses of collection centers (first model)

Collection center	Weighted loss
A	45.95
B	37.02
C	29.85
D	64.31

6.3 Evaluation Criteria for Second and Third Models

In evaluation of collection centers, one may have a scenario with three different categories of decision makers with multiple, conflicting, and incommensurate goals, as follows:

- *Consumers* whose primary concern is *convenience*
- *Local government officials* whose primary concern is *environmental consciousness*
- *Supply chain company executives* whose primary concern is *profit*

Therefore, the efficiency of a candidate collection center must be evaluated based on the maximized consensus among decision makers of the three categories. Sections 6.3.1, 6.3.2, and 6.3.3 present the lists of criteria that are considered for the three categories. Note that the criteria for the third category (supply chain company executives) are the same as those listed in section 6.2.1.

6.3.1 Criteria of Consumers

- Incentives from collection center (IC) (higher incentives imply higher motivation to participate)

- Distance from residential area (*DH*) (lower distance implies higher motivation to participate)
- Distance from roads (*DR*) (lower distance implies higher motivation to participate)
- Simplicity of collection process (*SP*) (simpler process implies higher motivation to participate)
- Employment opportunity (*EO*) (the higher, the better)
- Salary offered to employees at collection center (*SA*) (the higher, the better)

6.3.2   Criteria of Local Government Officials

- Distance from residential area (*DH*) (lower distance implies greater collection and hence lower disposal)
- Distance from roads (*DR*) (lower distance implies greater collection and hence lower disposal)

6.3.3   Criteria of Supply Chain Company Executives

- *n* value (the higher the *n* value, the lower the process defects, and hence the higher the profit; see section 4.16 for the concept of *n* Sigma)
- Per capita income of people in residential area (*PI*) (the higher it is, the more the number of resourceful discarded products, and the less the people will care about the incentives from the collection center)
- Space cost (*SC*) (the lower, the better)
- Labor cost (*LC*) (the lower, the better)
- Utilization of incentives from local government (*UI*) (the higher, the better)
- Distance from residential area (*DH*) (lower distance implies greater collection and hence greater profit)
- Distance from roads (*DR*) (lower distance implies greater collection and hence greater profit)
- Incentives from local government (*IG*) (higher incentives from local government imply higher incentives to consumers and hence greater collection)

6.4 Second Model (Eigen Vector Method, TOPSIS, and Borda's Choice Rule)

This model to select efficient collection centers is implemented in two phases. In the first phase, using the eigen vector method, impacts are given to the criteria identified for each category of decision makers (see section 6.3), and then TOPSIS (technique for order preference by similarity to ideal solution) is employed to find the efficiency of each candidate collection center, as evaluated by that category. In the second phase, Borda's choice rule is used to combine individual evaluations for each candidate collection center into a group evaluation or maximized consensus ranking. Furthermore, motivated by a paper [3] that addresses the necessity for building the strategic planning process "around a picture" to make the chairman of the concerned supply chain company easily understand the "dense documents filled with numbers" and to convince him that it is important to implement the proposed action (here, selection of particular collection centers), a GIS-based business mapping application, MapLand [4], is used to map the results obtained in both phases of this model.

The model to select efficient collection centers is presented using a numerical example. Three recovery facilities, E, F, and G, are considered for evaluation.

6.4.1 Phase I (Individual Decision Making)

Tables 6.5–6.7 show the pair-wise comparison matrices as formed for the consumers, local government officials, and supply chain company executives, respectively. Note that not all of the criteria are considered here, because the focus is on the methodology.

Tables 6.8–6.10 show the impacts given for the criteria of the consumers, local government officials, and supply chain company executives, respectively. These sets of impacts, calculated using the eigen vector method, are the elements of the normalized eigen vectors of pair-wise comparison matrices shown in tables 6.5–6.7, respectively. For example, the impacts of the criteria *DH* (distance from residential area) and *DR* (distance from roads), shown in table 6.9, are the elements of the normalized eigen vector of the pair-wise comparison matrix shown in table 6.6.

The decision matrices for the consumers, local government officials, and supply chain company executives, for implementation of the TOPSIS for each category of decision makers, are shown in tables 6.11–6.13, respectively. The elements of these matrices are the ranks (ranging from 1 to 10) assigned to the collection centers with respect to each criterion for evaluation. A lower rank implies higher efficiency (with respect to that criterion).

To facilitate the construction of the pair-wise comparison matrices (see tables 6.8–6.10) and the decision matrices (see tables 6.11–6.13), representatives from each category of decision makers could be invited to participate

TABLE 6.5

Pair-wise comparison matrix for consumers (second model)

Criteria	IC	DH	DR	SP	EO	SA
IC	1	1	1	2	1/2	1
DH	1	1	2	1	1	1/2
DR	1	1/2	1	1	1/7	1/2
SP	1/2	1	1	1	1	1
EO	2	1	7	1	1	1
SA	1	2	2	1	1	1

TABLE 6.6

Pair-wise comparison matrix for local government officials (second model)

Criteria	DH	DR
DH	1	2
DR	1/2	1

TABLE 6.7

Pair-wise comparison matrix for supply chain company executives (second model)

Criteria	PI	SC	LC	UI	DH	DR	IG
PI	1	2	4	6	8	9	4
SC	1/2	1	2	6	8	9	1
LC	1/4	1/2	1	1	1	1	1
UI	1/6	1/6	1	1	1	1	1
DH	1/8	1/8	1	1	1	1	1
DR	1/9	1/9	1	1	1	1	1
IG	1/4	1	1	1	1	1	1

TABLE 6.8

Impacts for consumers (second model)

Criteria	Impacts
IC	0.1621
DH	0.1515
DR	0.0960
SP	0.1434
EO	0.2533
SA	0.1938

TABLE 6.9

Impacts for local government officials (second model)

Criteria	Impacts
DH	0.6667
DR	0.3333

TABLE 6.10

Impacts for supply chain company executives (second model)

Criteria	Impacts
PI	0.3876
SC	0.2599
LC	0.0781
UI	0.0634
DH	0.0598
DR	0.0585
IG	0.0927

TABLE 6.11

Decision matrix for consumers (second model)

Collection centers	IC	DH	DR	SP	EO	SA
E	8	1	6	2	2	3
F	2	1	7	3	2	3
G	3	2	4	1	1	5

TABLE 6.12

Decision matrix for local government officials (second model)

Collection centers	DH	DR
E	2	1
F	2	1
G	3	2

TABLE 6.13

Decision matrix for supply chain company executives (second model)

Collection centers	PI	SC	LC	UI	DH	DR	IG
E	1	3	2	1	3	4	3
F	2	4	7	1	2	2	2
G	10	9	8	7	5	1	6

in relevant survey questionnaires, individual interviews, focus groups, and on-site observations.

Now one is ready to perform the six steps in the TOPSIS for each category of decision makers. The following steps show the implementation of the TOPSIS for the consumers to evaluate E, F, and G.

Step 1: *Construct the normalized decision matrix.* Table 6.14 shows the normalized decision matrix formed by applying equation (4.49) on each element of table 6.11 (decision matrix for the consumers). For example, the normalized rank of collection center F with respect to criterion *DH* (see tables 6.11 and 6.14) is calculated as follows:

$$r_{22} = \frac{1}{\sqrt{1^2 + 1^2 + 2^2}} = 0.4082.$$

Step 2: *Construct the weighted normalized decision matrix.* Table 6.15 shows the weighted normalized decision matrix for the consumers. This is constructed using the impacts of the criteria listed in table 6.8 and the normalized decision matrix in table 6.14. For example, the weighted normalized rank of collection center F with respect to criterion *DH*, i.e., 0.0619 (see table 6.15), is calculated by multiplying the impact of *DH*, i.e., 0.1515 (see table 6.8), with the normalized rank of F with respect to *DH*, i.e., 0.4082 (see table 6.14).

Step 3: *Determine the ideal and negative-ideal solutions.* Each column in the weighted normalized decision matrix shown in table 6.15 has a

TABLE 6.14

Normalized decision matrix for consumers (second model)

Collection centers	IC	DH	DR	SP	EO	SA
E	0.9117	0.4082	0.5970	0.5345	0.6667	0.4575
F	0.2279	0.4082	0.6965	0.8018	0.6667	0.4575
G	0.3419	0.8165	0.3980	0.2673	0.3333	0.7625

TABLE 6.15

Weighted normalized decision matrix for consumers (second model)

Collection centers	IC	DH	DR	SP	EO	SA
E	0.1478	0.0619	0.0573	0.0767	0.1689	0.0887
F	0.0370	0.0619	0.0669	0.1150	0.1689	0.0887
G	0.0554	0.1237	0.0382	0.0383	0.0844	0.1478

minimum rank and a maximum rank. They are the ideal and negative-ideal solutions, respectively, for the corresponding criterion.
For example (see table 6.15), with respect to criterion *DH*, the ideal
solution (minimum rank) is 0.0619, and the negative-ideal solution
(maximum rank) is 0.1237.

Step 4: *Calculate the separation distances.* The separation distances (see
table 6.16) for each collection center are calculated using equations
(4.52) and (4.53). For example, the positive separation distance for
collection center F (see table 6.16) is calculated using equation (4.52),
which contains the weighted normalized ranks of F (see table 6.15)
and the ideal solutions (obtained in step 3) for the criteria.

Step 5: *Calculate the relative closeness coefficient.* Using equation (4.54),
the relative closeness coefficient is calculated for each collection
center (see table 6.17). For example, the relative closeness coefficient
(i.e., 0.5435) for collection center F (see table 6.17) is the ratio of F's
negative separation distance (i.e., 0.1400) to the sum (i.e., 0.1400 +
0.1176 = 0.2576) of its negative and positive separation distances (see
table 6.16).

Step 6: *Form the preference order.* Because the best alternative is the one
with the highest relative closeness coefficient, the preference order
for the collection centers is G, F, and E (that means G is the best collection center, as evaluated by the consumers).

The TOPSIS is implemented for the local government officials and the supply chain company executives in a similar manner. The relative closeness
coefficients of the collection centers, as calculated for those two categories,
are shown in table 6.18.

Figure 6.2 shows the mapping classification, as created using MapLand
[4], for relative closeness coefficients. Figures 6.3–6.5 show the mapping of

TABLE 6.16

Separation distances for consumers (second model)

Collection centers	S*	S–
E	0.1458	0.0942
F	0.1176	0.1400
G	0.0875	0.1495

TABLE 6.17

Relative closeness coefficients for consumers (second model)

Collection centers	C*
E	0.3926
F	0.5435
G	0.6308

TABLE 6.18

Relative closeness coefficients for local government officials and supply chain company executives (second model)

Collection centers	Local government officials	Supply chain company executives
E	1.000	0.902
F	1.000	0.851
G	0.739	0.091

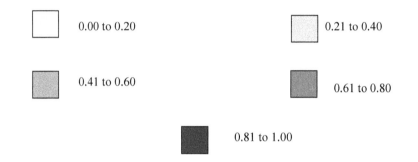

FIGURE 6.2

Relative closeness mapping classification (second model).

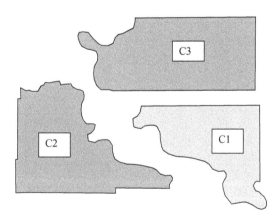

FIGURE 6.3

Mapping of collection centers for consumers (second model).

the relative closeness coefficients as calculated for the collection centers by the consumers, local government officials, and supply chain company executives, respectively.

It is assumed that each collection center is in one of the three regions, and the three regions are mapped with respect to the relative closeness coefficients. The darker the region is, the higher the efficiency the corresponding collection center has (with respect to the corresponding category of decision makers).

6.4.2 Phase II (Group Decision Making)

Table 6.19 shows the marks of the collection centers as given using Borda's choice rule for the consumers, local government officials, and supply chain company executives. Borda scores (group evaluations) calculated for E, F, and G (viz., 3, 5, and 3, respectively) are also shown. For example, the Borda score

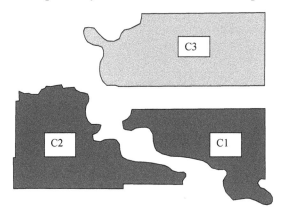

FIGURE 6.4
Mapping of collection centers for local government officials (second model).

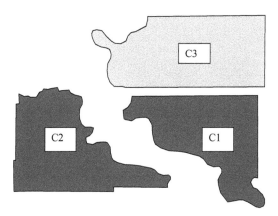

FIGURE 6.5
Mapping of collection centers for supply chain company executives (second model).

TABLE 6.19

Marks and Borda scores of collection centers (second model)

Collection centers	Consumers	Local government officials	Supply chain company executives	Borda scores
E	0	2	1	3
F	1	2	2	5
G	2	1	0	3

for F (i.e., 5) is calculated by summing the marks of F for consumers, local government officials, and supply chain company executives (i.e., 1 + 2 + 2). Because F has the highest Borda score, it is the best of the lot.

Figure 6.6 shows the mapping classification, as created using MapLand [4], for the Borda scores. Figure 6.7 shows the mapping of the Borda score of each collection center. It is obvious that F is the one with the highest efficiency.

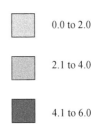

0.0 to 2.0

2.1 to 4.0

4.1 to 6.0

FIGURE 6.6
Borda score mapping classification (second model).

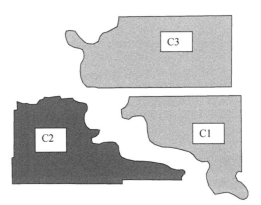

FIGURE 6.7
Mapping of collection centers with respect to Borda score (second model).

6.5 Third Model (Neural Networks, Fuzzy Logic, TOPSIS, Borda's Rule)

This model assumes that neither numerical nor linguistic impacts are available for the evaluation criteria. It employs a neural network [11] to evaluate the efficiency of a collection center of interest (which is being considered for inclusion in a reverse supply chain), using linguistic performance measures of collection centers that already exist in the reverse supply chain. To this end, the model to evaluate the efficiency of a collection center of interest is carried out in three phases. In the first phase, the ratings of existing collection centers are used to construct a neural network that, in turn, calculates impacts of criteria identified for each category of decision makers given in section 6.3. In the second phase, the impacts obtained in the first phase are used in a fuzzy TOPSIS (combination of fuzzy logic and TOPSIS) method to obtain the overall rating of the collection center of interest, as calculated for each category. Finally, in the third phase, Borda's choice rule is employed to calculate the maximized consensus rating (among the categories considered), i.e., efficiency, of the recovery facility of interest.

6.5.1 Phase I (Derivation of Impacts)

Suppose that one has the linguistic ratings of ten existing collection centers, as given by an expert in each category of decision makers described in section 6.3. Using fuzzy logic, these linguistic ratings are converted into triangular fuzzy numbers (TFNs). Table 6.20 shows not only one of the many ways for conversion of linguistic ratings into TFNs but also the defuzzified ratings of the corresponding TFNs. Tables 6.21–6.23 show the defuzzified overall rating of each existing collection center as well as the collection center's defuzzified rating with respect to each criterion, as evaluated by the consumers, local government officials, and supply chain company executives, respectively. Defuzzification of a TFN can be performed using equation (4.8).

A neural network is constructed and trained for each category of decision makers, using the defuzzified ratings of the existing collection centers with respect to criteria as input sets and the collection centers' defuzzified overall

TABLE 6.20

Conversion table for ratings (third model)

Linguistic ratings	TFNs	Defuzzified ratings
Very good (VG)	(7, 10, 10)	9
Good (G)	(5, 7, 10)	7.3
Fair (F)	(2, 5, 8)	5
Poor (P)	(1, 3, 5)	3
Very poor (VP)	(0, 0, 3)	1

TABLE 6.21

Ratings for consumers (third model)

Collection centers	IC	DH	DR	SP	EO	SA	Overall
C1	1	3	5	3	5	9	5
C2	9	1	3	5	7.3	9	7.3
C3	3	1	3	1	9	1	3
C4	3	9	1	7.3	1	7.3	5
C5	5	1	3	5	1	3	7.3
C6	9	3	7.3	3	5	7.3	3
C7	5	7.3	9	1	7.3	9	1
C8	1	5	1	5	3	1	9
C9	1	5	5	9	9	5	5
C10	5	9	5	3	9	3	1

TABLE 6.22

Ratings for local government officials (third model)

Collection centers	DH	DR	Overall
C1	1	3	5
C2	9	1	7.3
C3	3	1	3
C4	3	9	5
C5	5	1	7.3
C6	9	3	3
C7	5	7.3	1
C8	1	5	9
C9	1	5	5
C10	5	9	1

ratings as corresponding outputs. In the example, there are ten input–output pairs for each neural network because there are ten existing collection centers. Also, three layers are considered in each network, with five nodes in the hidden layer. The number of nodes in the output layer is one (for overall rating), and the number in the input layer is the number of criteria considered by the corresponding category. For example, figure 6.8 shows the neural network constructed and trained for the category of consumers.

After each neural network is trained, the following equation [2] is used to calculate the impacts of criteria considered by the corresponding category. Here, the absolute value of W_v is the impact of the vth input node upon the output node, n_V is the number of input nodes, n_H is the number of hidden nodes, I_{ij} is the connection weight from the ith input node to the jth hidden node, and O_j is the connection weight from the jth hidden node to the output node:

TABLE 6.23

Ratings for supply chain company executives (third model)

Collection centers	PI	SC	LC	UI	DH	DR	IG	Overall
C1	1	3	1	3	5	1	3	5
C2	9	1	3	7.3	3	5	7.3	7.3
C3	3	1	7.3	9	1	7.3	9	3
C4	3	9	5	1	5	3	1	5
C5	5	1	5	5	9	9	5	7.3
C6	9	3	9	5	3	9	3	3
C7	5	7.3	3	1	7.3	9	1	1
C8	1	5	1	3	1	3	5	9
C9	1	5	3	5	9	9	1	5
C10	5	9	1	3	5	7.3	7.3	1

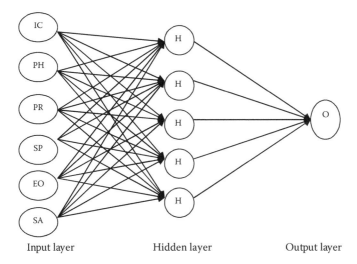

FIGURE 6.8

Neural network for consumers (third model).

$$|W_v| = \frac{\sum_{j}^{n_H} \dfrac{I_{vj}}{\sum_{i}^{n_V} |I_{ij}|} O_j}{\left(\sum_{i}^{n_V} \left| \sum_{j}^{n_H} \dfrac{I_{vj}}{\sum_{i}^{n_V} |I_{ij}|} O_j \right| \right)}$$

(6.2)

TABLE 6.24

Impacts for consumers (third model)

Criteria	IC	DH	DR	SP	EO	SA
Impacts	0.01	0.13	0.06	0.18	0.19	0.43

TABLE 6.25

Impacts for local government officials (third model)

Criteria	DH	DR
Impacts	0.33	0.67

TABLE 6.26

Impacts for supply chain company executives (third model)

Criteria	PI	SC	LC	UI	DH	DR	IG
Impacts	0.24	0.09	0	0.18	0.25	0.1	0.13

Tables 6.24–6.26 show the impacts of the criteria considered for the consumers, local government officials, and supply chain company executives, respectively.

6.5.2   Phase II (Individual Decision Making)

Suppose that there are three collection centers, C11, C12, and C13, of interest. A fuzzy TOPSIS method uses the impacts obtained in the first phase to calculate the overall ratings of the three collection centers.

The decision matrices formed for the consumers, local government officials, and supply chain company executives (with defuzzified ratings for C11, C12, and C13) in this example are shown in tables 6.27–6.29, respectively (Table 6.20 is used here, as well, to convert linguistic ratings given by each category into TFNs.) As in the second model (see section 6.3), the matrices can be constructed by inviting representatives from each category of decision makers to participate in relevant survey questionnaires, individual interviews, focus groups, and on-site observations.

Now one is ready to perform the six steps in the TOPSIS for each category of decision makers. The following steps show the implementation of the TOPSIS for the consumers, to evaluate C11, C12, and C13.

Step 1: *Construct the normalized decision matrix.* Table 6.30 shows the normalized decision matrix formed by applying equation (4.49) on each element of table 6.27 (decision matrix for the consumers). For example, the normalized rating of collection center C12 with respect to criterion *DH* (see tables 6.27 and 6.30) is calculated as follows:

TABLE 6.27

Decision matrix for consumers (third model)

Collection centers	IC	DH	DR	SP	EO	SA
C11	3	9	1	7.33	1	7.33
C12	5	1	3	5	1	3
C13	9	3	7.33	3	5	7.33

TABLE 6.28

Decision matrix for local government officials (third model)

Collection centers	DH	DR
C11	3	9
C12	5	1
C13	9	3

TABLE 6.29

Decision matrix for supply chain company executives (third model)

Collection centers	PI	SC	LC	UI	DH	DR	IG
C11	5	1	5	5	9	9	5
C12	9	3	9	5	3	9	3
C13	5	7.33	3	1	7.33	9	1

TABLE 6.30

Normalized decision matrix for consumers (third model)

Collection centers	IC	DH	DR	SP	EO	SA
C11	0.278	0.943	0.125	0.783	0.192	0.679
C12	0.466	0.105	0.376	0.534	0.192	0.278
C13	0.839	0.314	0.918	0.320	0.962	0.679

$$r_{22} = \frac{1}{\sqrt{9^2 + 1^2 + 3^2}} = 0.105.$$

Step 2: *Construct the weighted normalized decision matrix.* Table 6.31 shows the weighted normalized decision matrix for the consumers. This is constructed using the impacts of the criteria listed in table 6.24 and the normalized decision matrix in table 6.30. For example, the

TABLE 6.31

Weighted normalized decision matrix for consumers (third model)

Collection centers	IC	DH	DR	SP	EO	SA
C11	0.004	0.122	0.007	0.140	0.036	0.294
C12	0.006	0.014	0.022	0.095	0.036	0.120
C13	0.011	0.041	0.053	0.057	0.181	0.294

weighted normalized rank of collection center C12 with respect to criterion *DH*, i.e., 0.014 (see table 6.31), is calculated by multiplying the impact of *DH*, i.e., 0.129 (see table 6.24), with the normalized rating of C12 with respect to *DH*, i.e., 0.105 (see table 6.30).

Step 3: *Determine the ideal and negative-ideal solutions.* Each column in the weighted normalized decision matrix shown in table 6.31 has a maximum rating and a minimum rating. They are the ideal and negative-ideal solutions, respectively, for the corresponding criterion. For example (see table 6.31), with respect to criterion *DH*, the ideal solution (maximum rating) is 0.122, and the negative-ideal solution (minimum rating) is 0.014.

Step 4: *Calculate the separation distances.* The separation distances (see table 6.32) for each collection center are calculated using equations (4.52) and (4.53). For example, the positive separation distance for collection center C12 (see table 6.32) is calculated using equation (4.52), which contains the weighted normalized ratings of C12 (see table 6.31) and the ideal solutions (obtained in step 3) for the criteria.

Step 5: *Calculate the relative closeness coefficient.* Using equation (4.54), the relative closeness coefficient for each collection center is calculated (see table 6.33). For example, the relative closeness coefficient (i.e., 0.137) for collection center C12 (see table 6.33) is the ratio of C12's negative separation distance (i.e., 0.041) to the sum (i.e., 0.041 + 0.257 = 0.298) of its negative and positive separation distances (see table 6.32).

Step 6: *Rank the preference order.* Because the best alternative is the one with the highest relative closeness coefficient, the preference order

TABLE 6.32

Separation distances for consumers (third model)

Collection centers	S*	S–
C11	0.152	0.221
C12	0.257	0.041
C13	0.116	0.233

TABLE 6.33

Relative closeness coefficients for consumers (third model)

Collection centers	$C*$
C11	0.592
C12	0.137
C13	0.668

for the collection centers is C13, C11, and C12 (that means C13 is the best collection center, as evaluated by the consumers).

The TOPSIS is implemented for the local government officials and the supply chain company executives in a similar manner. The relative closeness coefficients of the collection centers, as calculated for those two categories, are shown in table 6.34.

6.5.3 Phase III (Group Decision Making)

Table 6.35 shows the marks of the collection centers as given using Borda's choice rule for the consumers, local government officials, and supply chain company executives. Borda scores (group evaluations) calculated for C11, C12, and C13 (viz., 5, 1, and 3, respectively) are also shown. For example, the Borda score for C12 (i.e., 1) is calculated by summing the marks of C12 for the consumers, local government officials, and supply chain company executives (i.e., 0 + 0 + 1). Because C11 has the highest Borda score, it is the best of the lot.

TABLE 6.34

Relative closeness coefficients for local government officials and supply chain company executives (third model)

Collection centers	Local government officials	Supply chain company executives
C11	0.754	0.619
C12	0.096	0.502
C13	0.354	0.415

TABLE 6.35

Marks and Borda scores of collection centers (third model)

Collection centers	Consumers	Local government officials	Supply chain company executives	Borda scores
C11	1	2	2	5
C12	0	0	1	1
C13	2	1	0	3

6.6 Fourth Model (ANP and Goal Programming)

In this model, first the analytic network process (ANP) is used to calculate the performance indices (efficiency scores) of candidate collection centers, with respect to qualitative criteria taken from the perspective of a remanufacturing facility interested in buying used products from the collection centers. Then goal programming is employed to determine the quantities of used products to be transported from the candidate collection centers to the remanufacturing facility while satisfying two important goals of the remanufacturing facility: to maximize total value of purchase and minimize total cost of purchase. Sections 6.6.1 and 6.6.2 present the applications of ANP and goal programming, respectively.

6.6.1 Application of ANP

The problem of evaluating the efficiencies of the candidate collection centers is framed as a four-level hierarchy (see figure 6.9). The first level contains the objective of evaluation of the candidate collection centers. The second level consists of the main evaluation criteria taken from the perspective of a remanufacturing facility. The third level contains the subcriteria under each

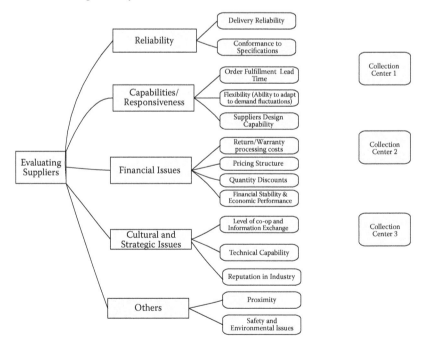

FIGURE 6.9
Hierarchical structure for ANP (fourth model).

main criterion. The fourth level contains the candidate collection centers. The main and subcriteria considered are the following (see [5–9]):

- *Reliability*: This criterion relates to a collection center delivering the used products at the right time, at the right remanufacturing facility, in the right quantity, and in the promised condition. The subcriteria considered under this main criterion are (1) delivery reliability and (2) conformance to standards.

- *Capability/responsiveness*: This criterion reflects the velocity at which a collection center supplies the used products to the remanufacturing facility and the collection center's ability to adapt to sudden demand fluctuations. The subcriteria considered here are (1) order fulfillment lead time, (2) flexibility in adapting to demand fluctuations, and (3) design capabilities.

- *Financial issues*: This criterion reflects the costs and other financial aspects involved. The subcriteria are (1) return/warranty processing costs, (2) pricing structure, (3) quantity discounts, and (4) financial stability and economic performance of the collection center.

- *Cultural and strategic issues*: This criterion consists of the following subcriteria: (1) level of cooperation and information exchange between the collection center and the remanufacturing facility, (2) the collection center's reputation in the industry, and (3) the collection center's technical capability (how knowledgeable the collection center is about the product).

- *Others*: This criterion considers miscellaneous aspects that are not considered in the other criteria. These aspects are (1) proximity of the collection center to the remanufacturing facility (it affects the transportation cost and the transit time) and (2) safety and environmental aspects (because the collection center and the remanufacturing facility are closely involved, any safety issues with the collection center directly reflect on the remanufacturing facility's reputation; environmental aspects are concerned about the collection center's effort in pursuing environmental consciousness or a "green" image).

It is assumed that there exist interdependencies among the subcriteria on the third level in the hierarchy.

Three collection centers, S1, S2, and S3, are considered in the numerical example to illustrate the application of ANP. Table 6.36 shows the pair-wise comparison matrix for the main criteria (second level in the hierarchy) and also the normalized eigen vector of the matrix. The elements of the normalized eigen vector are the impacts given to the main criteria with respect to the objective (first level in the hierarchy).

Tables 6.37–6.41 show the pair-wise comparison matrices of subcriteria with respect to their main criteria and also the corresponding normalized eigen vectors of the matrices. Note that each of the matrices in tables 6.36–6.41 has

TABLE 6.36
Comparative importance values of main criteria (fourth model)

Criteria	Reliability	Responsiveness	Financial issues	Cultural and strategic issues	Others	Normalized eigen vector
Reliability	1	1/4	3	1	1/6	0.114
Responsiveness	4	1	1	5	3	0.431
Financial issues	1/3	1/5	1	1/2	1/3	0.065
Cultural and strategic issues	1	1/3	2	1	1/3	0.102
Others	6	1/3	3	3	1	0.285

TABLE 6.37
Comparative importance values of subcriteria under reliability (fourth model)

Subcriteria	Delivery reliability	Conformance to specs	Normalized eigen vector
Delivery reliability (DR)	1	1/3	0.25
Conformance to specs (CS)	3	1	0.75

TABLE 6.38
Comparative importance values of subcriteria under responsiveness (fourth model)

Subcriteria	Order fulfillment lead time (OFT)	Flexibility (F)	Design capability (DC)	Normalized eigen vector
Order fulfillment lead time (OFT)	1	3	4	0.623
Flexibility (F)	1/3	1	2	0.239
Design capability (DC)	1/4	1/2	1	0.137

a consistency ratio CR less than or equal to 0.1. For example, for the matrix in table 6.39, CR, using equation (4.1), is

$$\frac{(\lambda_{max}-n)}{(n-1)(R)} = \frac{4.01037-4}{(4-1)\times0.90} = 0.0038.$$

Table 6.42 shows the matrix of interdependencies (called the super matrix M) among the subcriteria with respect to their main criteria. This super matrix M is made to converge to obtain a long-term stable set of impacts. For convergence, M must be made column stochastic, which is done by raising M to the power of 2^{k+1}, where k is an arbitrarily large number. In the example, k = 59. Table 6.43 shows the converged super matrix.

TABLE 6.39

Comparative importance values of subcriteria under "financial issues" (fourth model)

Subcriteria	Returns/warranty processing costs (RC)	Pricing structure (PS)	Quantity discounts (QD)	Economic performance and financial stability (EP)	Normalized eigen vector
Returns/warranty processing costs (RC)	1	1/3	1/2	1/4	0.099
Pricing structure (PS)	3	1	2	1	0.345
Quantity discounts (QD)	2	1/2	1	1/2	0.185
Economic performance and financial stability (EP)	4	1	2	1	0.37

TABLE 6.40

Comparative importance values of subcriteria under cultural and strategic issues (fourth model)

Subcriteria	Level of co-op and information exchange (Co-op)	Technical capability (TC)	Reputation	Normalized eigen vector
Level of co-op and information exchange (Co-op)	1	5	6	0.722
Technical capability (TC)	1/5	1	2	0.174
Reputation (R)	1/6	1/2	1	0.103

TABLE 6.41

Comparative importance values of subcriteria under others (fourth model)

Subcriteria	Proximity (P)	Safety and environment (SE)	Normalized eigen vector
Proximity (P)	1	1/3	0.25
Safety and environment (SE)	3	1	0.75

TABLE 6.42

Matrix of interdependencies (super matrix M) (fourth model)

	DR	CS	OFT	F	DC	RC	PS	QD	EP	Co-op	TC	R	P	SE
DR	0	1	0	0	0	0	0	0	0	0	0	0	0	0
CS	1	0	0	0	0	0	0	0	0	0	0	0	0	0
OFT	0	0	0	0.8	0.75	0	0	0	0	0	0	0	0	0
F	0	0	0.75	0	0.25	0	0	0	0	0	0	0	0	0
DC	0	0	0.25	0.2	0	0	0	0	0	0	0		0	0
RC	0	0	0	0	0	0	0.29	0.13	0.201	0	0	0	0	0
PS	0	0	0	0	0	0.53	0	0.62	0.6	0	0	0	0	0
QD	0	0	0	0	0	0.29	0.16	0	0.11	0	0	0	0	0
EP	0	0	0	0	0	0.163	0.53	0.23	0	0	0	0	0	0
Co-op	0	0	0	0	0	0	0	0	0	0	0.75	0.83	0	0
TC	0	0	0	0	0	0	0	0	0	0.8	0	0.16	0	0
R	0	0	0	0	0	0	0	0	0	0.2	0.25	0	0	0
P	0	0	0	0	0	0	0	0	0	0	0	0	0	1
SE	0	0	0	0	0	0	0	0	0	0	0	0	1	0

TABLE 6.43

Converged super matrix (fourth model)

Subcriteria	Stabilized relative impact
Delivery reliability	1
Conformance to specs	1
Order fulfillment LT	0.44
Flexibility	0.38
Design capability	0.18
Returns/warranty	0.19
Pricing	0.38
Quantity discounts	0.15
Stability and economic performance	0.27
Co-op and information exchange	0.44
Technical capability	0.38
Reputation	0.18
Proximity	1
Safety and environment	1

Table 6.44 shows the relative ratings of the candidate collection centers, S1, S2, and S3, with respect to the subcriteria. These ratings are obtained after carrying out pair-wise comparisons between the candidate collection centers with respect to the subcriteria and then obtaining the normalized eigen vector.

To obtain pair-wise comparisons, interdependencies, and relative ratings (see tables 6.37–6.44), decision makers could be invited to participate in relevant survey questionnaires, individual interviews, focus groups, and on-site observations.

Table 6.45 shows the desirability index calculated for each candidate collection center using equation (4.2).

The overall performance index for each of the three collection centers is calculated by multiplying the desirability index (see table 6.45) of each collection center for each main criterion by the impact of that criterion (see

TABLE 6.44

Relative ratings of collection centers with respect to subcriteria (fourth model)

Subcriteria/alternate collection centers	S1	S2	S3
DR	0.33	0.141	0.524
CS	0.345	0.543	0.11
OFT	0.274	0.068	0.657
F	0.109	0.309	0.581
DC	0.09	0.25	0.652
RC	0.681	0.216	0.102
PS	0.309	0.581	0.109
QD	0.376	0.151	0.471
EP	0.137	0.623	0.239
Co-op	0.33	0.075	0.59
TC	0.137	0.239	0.623
R	0.67	0.23	0.12
P	0.292	0.092	0.615
SE	0.137	0.239	0.623

TABLE 6.45

Desirability indices (fourth model)

Criteria/collection centers	S1	S2	S3
Reliability	0.3429	0.4432	0.2138
Responsiveness	0.0874	0.0528	0.2488
Financial issues	0.0783	0.148551	0.054
Cultural and strategic issues	0.1271	0.0438	0.2299
Others	0.176	0.2027	0.6211

TABLE 6.46

Overall performance indices (fourth model)

Collection center	Performance index
S1	0.2318
S2	0.2322
S3	0.5359

table 6.36) and summing over all the criteria. Table 6.46 shows the overall performance indices (efficiencies) for the three collection centers.

6.6.2   Application of Goal Programming

This section presents the application of goal programming to determine the quantities of used products to be transported from the candidate collection centers to a remanufacturing facility of interest while satisfying two important goals of the remanufacturing facility: to maximize total value of purchase and minimize total cost of purchase. Section 6.6.2.1 gives the nomenclature used in the methodology, and section 6.6.2.2 presents the problem formulation and a numerical example.

6.6.2.1   Nomenclature for Problem Formulation

c_i Unit purchasing cost of used product at collection center i

d_j Demand for used product j

g Goal index

i Collection center index, $i = 1, 2, ..., s$

k_i Capacity of collection center i

p_i Probability of breakage of used products purchased from collection center i

p_{max} Maximum allowable probability of breakage

Q_i Decision variable representing the quantity to be purchased from collection center i

s Number of candidate collection centers

w_i Performance index of collection center i obtained by carrying out ANP

6.6.2.2   Problem Formulation

The following two goals of the remanufacturing facility are considered:

1. Maximize the total value of purchase (TVP).
2. Minimize the total cost of purchase (TCP).

Whereas the first goal involves minimizing the underachievement of the target, the second goal involves minimizing the overachievement of the target. It is at the discretion of the decision maker to add any other goals that are relevant to the situation.

$$\text{Goal 1: } \textit{Maximize TVP: } \sum_{i=1}^{s} w_i \times Q_i \tag{6.3}$$

$$\text{Goal 2: } \textit{Maximize TCP: } \sum c_i \times Q_i \tag{6.4}$$

$$\textit{Capacity constraint: } Q_i \leq k_i \tag{6.5}$$

$$\textit{Demand constraint: } \sum_i Q_i = d_j \tag{6.6}$$

$$\textit{Quality constraint: } d_j \times p_{\max} \geq \sum_{i=1}^{s} Q_i \times p_i \tag{6.7}$$

$$\textit{Nonnegativity constraint: } Q_i \geq 0 \tag{6.8}$$

The three candidate collection centers from section 6.6.1 are considered in this numerical example as well. Table 6.47 shows the data used for the goal programming problem (note that only one used product type is considered), and table 6.48 shows the results obtained by solving the problem using LINGO (v4).

From the application of ANP (see section 6.6.1), S3 is the highest-ranked collection center. When no other system constraints are in place, 750 units might be ordered from S3 before considering other collection centers. However, from the results obtained from the application of goal programming, it can be noticed that 650 units are ordered from S2 and the remaining 350 units are ordered from S3. This may be attributed to the fact that the unit purchasing cost at S3 is higher than that at S2 (this was not considered in the application of ANP, where collection centers were evaluated with respect to qualitative criteria).

The total cost of purchase is found to be $935, and the total value of purchase is 338.54 (aspiration level = 250).

TABLE 6.47

Data for goal programming model (fourth model)

Collection center	S1	S2	S3
Capacity	300	650	750
Unit purchasing cost	1.2	0.9	1.0
Breakage probability	0.03	0.015	0.01
Net demand for the product = 1,000			
Maximum acceptable breakage probability = 0.025			

TABLE 6.48

Results (fourth model)

Collection center	ANP rating	Quantity ordered
S1	0.2318	0
S2	0.2322	650
S3	0.5359	350
Total value of purchase (TVP) = 338.54		
Total cost of purchase (TCP) = 935		

6.7 Fifth Model (Eigen Vector Method, Taguchi Loss Function, and Goal Programming)

In this model, first the eigen vector method and Taguchi loss function are used to calculate the weighted Taguchi losses (inefficiency scores) of candidate collection centers, with respect to qualitative criteria taken from the perspective of a remanufacturing facility interested in buying used products from the collection centers. Then goal programming is employed to determine the quantities of used products to be transported from the candidate collection centers to the remanufacturing facility while satisfying two important goals of the remanufacturing facility: to minimize total loss of profit and minimize total cost of purchase. Section 6.7.1 presents the application of the eigen vector method and Taguchi loss function. Then section 6.7.2 presents the application of goal programming.

6.7.1 Application of Eigen Vector Method and Taguchi Loss Function

The following four qualitative criteria (taken from the perspective of a remanufacturing facility) are considered to evaluate the candidate collection centers:

- Quality of used products (the smaller the defect rate of the used products supplied, the better)

- On-time delivery (the smaller the number of delayed deliveries, the better)
- Proximity (the closer the collection center from the remanufacturing facility, the better)
- Cultural and strategic issues, such as flexibility in adapting to demand fluctuations, level of cooperation and information exchange, green image, and financial stability/economic performance (see section 6.6.1 for explanations of these issues)

For the numerical example, three candidate collection centers, S4, S5, and S6, are considered. Table 6.49 shows the pair-wise comparison matrix for the criteria* and also the normalized eigen vector of the matrix. The elements of the normalized eigen vector represent the impacts given to the criteria. Table 6.49 has a consistency ratio CR of less than 0.1. Specifically, it is

$$\frac{(\lambda_{max}-n)}{(n-1)(R)} = \frac{4.13-4}{(4-1)\times 0.90} = 0.048.$$

It could be difficult to quantify the cultural and strategic issues criterion for the calculation of Taguchi losses. To this end, Monczka and Trecha's [10] service factor rating (SFR) is used. SFR includes performance factors difficult to quantify but decisive in the selection process. In practice, experts rate these performance factors. For a given collection center, these ratings on all factors are summed and averaged to obtain a total service rating. The collection center's service factor percentage is obtained by dividing the total service rating by the total number of points possible. Table 6.50 shows the service factor ratings for the various aspects of the cultural and strategic issues criterion. The ratings are given on a scale of 1–10, the level of performance being directly proportional to the rating.

TABLE 6.49

Comparative importance values of criteria (fifth model)

Criteria	Quality	On-time delivery	Proximity	Cultural and strategic issues	Normalized eigen vector
Quality	1	3	7	1	0.384899
On-time delivery	0.33	1	4	0.2	0.137363
Proximity	0.14	0.25	1	0.166	0.052674
Cultural and strategic issues	1	5	6	1	0.425064

---

* Only the main criteria are considered in this model.

TABLE 6.50

Service factor ratings for cultural and strategic issues (fifth model)

Collection center	Flexibility	Level of co-op and information exchange	Green image	Financial stability and economic performance	Average	Average ÷ 10
S4	7	6	6	4	5.75	57.5%
S5	5	7	8	5	6.25	62.5%
S6	6	5	8	8	6.75	67.5%

TABLE 6.51

Decision variables for selecting collection centers (fifth model)

Criteria	Target value	Range	Specification limit
Quality	0%	0–30%	30%
On-time delivery	0	10–0–5	10 days earlier, 5 days delay
Proximity	Closest	0–40%	40%
Cultural and strategic issues	100%	100–50%	50%

Table 6.51 shows the decision variables for calculating the Taguchi losses for the three collection centers.

Consider the quality criterion. The target defect rate/breakage probability is zero, at which there is no loss to the remanufacturing facility, and the upper specification limit for the defect rate/breakage probability is 30%, at which there is 100% loss to the manufacturer. For on-time delivery, the remanufacturing facility will incur losses if the products are delivered late and before the scheduled requirement. The specification limit of delivery delay is 5 days, and early delivery is 10 days, meaning that the remanufacturing facility will incur 100% loss if the deliveries are delayed by 5 days and are delivered 10 days before the scheduled delivery date. For proximity, loss will be zero at the closest collection center and the specification limit is up to 40% of the closest collection center, meaning that the remanufacturing facility will incur 100% loss when the collection center's distance reaches the specification limit. For cultural and strategic issues, the specification limit is 50% for the service factor percentage, at which the loss will be 100%, whereas there will be no loss incurred at a service factor percentage of 100%. The values of the loss coefficient, k (see equations 4.66–4.68), are 1111.11, 625, and 25 for quality, proximity, and cultural and strategic issues, respectively. For on-time delivery, $k_1 = 4$ and $k_2 = 1$ (because an unequal two-sided specification limit is considered for on-time delivery, there exist two loss coefficients, k_1 and k_2).

Table 6.52 shows the characteristic value and the relative value of each criterion for the three collection centers, S4, S5, and S6. For S4, the quality

TABLE 6.52

Characteristic and relative values of criteria (fifth model)

Collection center	Quality		On-time delivery		Proximity		Cultural and strategic issues	
	Value	Relative value	Value	Relative value	Value	Relative value	Value	Relative value
S4	15%	15%	+3	+3	8	33.33%	57.5%	57.5%
S5	20%	20%	+1	+1	6	0	62.5%	62.5%
S6	10%	10%	−8	−8	9	50%	67.5%	67.5%

characteristic value is 15% defect rate, which translates to 15% deviation from the target value. The relative values, together with the value of the loss coefficient, k, are used to calculate (see equations (4.66)–(4.68)) the Taguchi loss for each collection center for each criterion (see table 6.53).

The weighted Taguchi loss (sum product of criteria impacts and Taguchi losses) is then calculated from the Taguchi losses of the collection centers (see table 6.53) and the impacts of the evaluation criteria (see table 6.49). Table 6.54 shows the weighted Taguchi loss and the normalized Taguchi loss for each collection center.

6.7.2 Application of Goal Programming

This section presents the application of goal programming to determine the quantities of used products to be transported from the candidate collection centers to a remanufacturing facility of interest while satisfying two important goals of the remanufacturing facility: to minimize total loss of profit and minimize total cost of purchase. Section 6.7.2.1 gives the nomenclature

TABLE 6.53

Taguchi losses (fifth model)

Collection center	Quality	On-time delivery	Proximity	Cultural and strategic issues
S4	24.99	36	69.43	75.61
S5	44.44	4	0	64
S6	11.11	64	156.25	54.86

TABLE 6.54

Weighted Taguchi losses (fifth model)

Collection center	Weighted Taguchi loss	Normalized Taguchi loss
S4	50.37	0.36
S5	44.86	0.32
S6	44.62	0.32

used in the methodology, and section 6.7.2.2 presents the problem formulation and a numerical example.

6.7.2.1 Nomenclature Used in the Methodology

B_j Budget allocated for collection center j

c_j Unit purchasing cost of used product at collection center j

d_k Demand for used product k

g Goal index

j Collection center index, $j = 1, 2, ..., s$

$Loss_j$ Total loss of collection center j for all the evaluation criteria

r_j Capacity of collection center j

p_j Probability of breakage of used products purchased from collection center j

p_{max} Maximum allowable probability of breakage

Q_j Decision variable representing the purchasing quantity from collection center j

s Number of candidate collection centers

w_i Impact of criterion i calculated by the eigen vector method

X_{ij} Taguchi loss of collection center j for criterion i

6.7.2.2 Problem Formulation

The following two goals of the remanufacturing facility are considered:

1. Minimize the total loss of profit (TLP).
2. Minimize the total cost of purchase (TCP).

It is at the discretion of the decision maker to add any other goals that are relevant to the situation.

$$\text{Goal 1: } Minimize\ TLP: \sum_{j=1}^{s} Loss_j \times Q_j = TLP \qquad (6.9)$$

$$\text{Goal 2: } Minimize\ TCP: \sum_{j=1}^{s} c_j * Q_j = TCP \qquad (6.10)$$

$$\text{Capacity constraint: } Q_i \leq r_i \tag{6.11}$$

$$\text{Demand constraint: } \sum_j Q_j = d_j \tag{6.12}$$

$$\text{Quality constraint: } d_k \times p_{max} \geq \sum_{j=1}^{s} Q_j \times p_j \tag{6.13}$$

$$\text{Nonnegativity constraint: } Q_j \geq 0 \tag{6.14}$$

The three candidate collection centers from section 6.7.1 are considered in this numerical example as well. Table 6.55 shows the data used for the goal programming problem (note that only one used product type is considered), and table 6.56 shows the results obtained by solving the problem using LINGO (v4).

From the application of eigen vector method and Taguchi loss function (see section 6.7.1), S6 is the highest-ranked collection center. When no other system constraints are in place, 750 units might be ordered from S6 before considering other collection centers. However, from the results obtained from the application of goal programming, it can be noticed that 543 units are ordered from S5 and the remaining 457 units are ordered from S6. This may be attributed to

TABLE 6.55

Data for goal programming model (fifth model)

Collection center	S4	S5	S6
Capacity	300	650	750
Unit purchasing cost	1.2	0.9	1.0
Breakage probability	0.03	0.015	0.01
Net demand for the product = 1,000			
Maximum acceptable breakage probability = 0.025			

TABLE 6.56

Results (fifth model)

Collection center	Normalized Taguchi loss	Quantity ordered
S4	0.360148	0
S5	0.32078	543
S6	0.319072	457
Total loss of purchase (TLP) = 320		
Total cost of purchase (TCP) = 945.7		

the fact (besides the system constraints considered in the goal programming problem) that the unit purchasing cost at S6 is higher than that at S5.

6.8   Conclusions

In this chapter, five models are presented for identifying efficient collection centers in a region where a reverse supply chain is to be designed. The first model employs eigen vector method and Taguchi loss function. The second model uses eigen vector method, technique for order preference by similarity to ideal solution (TOPSIS), and Borda's choice rule. The third model employs neural networks, fuzzy logic, TOPSIS, and Borda's choice rule. The fourth model uses analytic network process (ANP) and goal programming. The fifth model applies eigen vector method, Taguchi loss function, and goal programming.

References

1. Krajewski, L. J., and Ritzman, L. P. 2004. *Operations management: Processes and value chains*, chap. 10. 7th ed. Englewood Cliffs, NJ: Pearson Prentice Hall.
2. Cha, Y., and Jung, M. 2003. Satisfaction assessment of multi-objective schedules using neural fuzzy methodology. *International Journal of Production Research* 41:1831–49.
3. Kim, C. W., and Mauborgne, R. 2002. Charting your company's future. *Harvard Business Review*, June, pp. 77–82.
4. http://www.softwareillustrated.com/mapland.html.
5. Choi, T. Y., and Hartley, J. L. 1996. An exploration of supplier selection practices across the supply chain. *Journal of Operations Management* 14:333–43.
6. Kannan, V. R., and Tan, K. C. 2002. Supplier selection and assessment: Their impact on business performance. *Journal of Supply Chain Management* 38:11–21.
7. Lee, E. K., Ha, S., and Kim, S. K. 2001. Supplier selection and management system considering relationships in SCM. *IEEE Transactions on Engineering Management* 48:307–18.
8. Muralidharan, C., Anantharaman, N., and Deshmukh, S. G. 2002. A multi-criteria group decision making model for supplier rating. *Journal of Supply Chain Management* 38:22–33.
9. Weber, C. A., Current, J. R., and Benton, W. C. 1991. Vendor selection criteria and methods. *European Journal of Operational Research* 50:2–18.
10. Monczka, R. M., and Trecha, S. J. 1998. Cost-based supplier performance evaluation. *Journal of Purchasing and Material Management* 2–7.
11. Haykin, S. 1998. *Neural networks: A comprehensive foundation.* 2nd ed. Englewood Cliffs, NJ: Prentice Hall.

7

Evaluation of Recovery Facilities

7.1 The Issue

In addition to selecting efficient collection centers (see chapter 6), strategic planning of a reverse supply chain involves selecting efficient recovery facilities where reprocessing operations, such as disassembly and recycling/remanufacturing, are carried out. The various scenarios for evaluating recovery facilities for efficiency could differ as follows:

1. Evaluation criteria could be given numerical impacts* (importance values).
2. Evaluation criteria could be presented in terms of ranges of different degrees of desirability.
3. Decision makers could have conflicting criteria for evaluation, and impacts for evaluation criteria are given.
4. Decision makers could have conflicting criteria for evaluation, and impacts for evaluation criteria are not given (hence must be derived).
5. A very simple evaluation technique could be desired (where only the "most important" evaluation criteria are considered).

In this chapter, various models to address the above scenarios are presented. The first model addresses scenario 1 and employs the analytic hierarchy process. The second model is for scenario 2 and employs linear physical programming. The third and fourth models are for scenarios 3 and 4, respectively. The main difference between these two models lies in the way the impacts are assigned to the criteria for evaluation of recovery facilities. Whereas the third model uses the eigen vector method, technique for order preference by similarity to ideal solution (TOPSIS), and Borda's choice rule, the fourth model uses neural networks, fuzzy logic, TOPSIS, and Borda's

---

* In this chapter, we use the term *impacts* instead of the conventional term *weights* for importance values of evaluation criteria. This is to avoid confusion in the presence of the connection weights between different nodes of the neural network used in section 7.6.

choice rule. The fifth model is for scenario 5 and uses a simple two-dimensional chart to identify efficient recovery facilities.

This chapter is organized as follows: Section 7.2 presents the first model. Section 7.3 shows the second model. Section 7.4 presents the different decision makers (and their criteria) that are considered in the third and fourth models. Section 7.5 presents the third model. Section 7.6 gives the fourth model. Section 7.7 presents the fifth model. Finally, section 7.8 gives some conclusions.

7.2   First Model (Analytic Hierarchy Process)

In this section, analytic hierarchy process (AHP) is employed to identify efficient facilities from a set of candidate recovery facilities operating in a region where a reverse supply chain is to be designed. Section 7.2.1 presents the three-level hierarchy for the AHP model, and section 7.2.2 gives a numerical example for the model.

7.2.1   Three-Level Hierarchy

The first level in the hierarchy contains the primary objective, i.e., to identify efficient facilities from a set of candidate recovery facilities. The last level in the hierarchy contains the candidate recovery facilities. The level in the middle contains criteria that must somehow be useful in comparing the candidate recovery facilities. For example, one of the criteria used in the hierarchy (see figure 7.1) is the fixed cost of the facility (CO). This criterion can compare the candidate facilities on the third level.

Though the criteria to be considered in a reverse supply chain seem similar to those considered in a forward supply chain (for example, [3] and [4]), there are three special factors in a reverse supply chain, that need to be incorporated in AHP in such a way that the hierarchy levels are not disturbed.

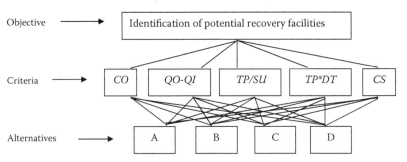

FIGURE 7.1
Three-level hierarchy for AHP (first model).

Those special factors are average quality of used products, average supply of used products, and average disassembly time of used products.

- *Average quality of used products*: Unlike in a forward supply chain, components of incoming goods (used products) of even the same type in a recovery facility are likely to be of varied quality (worn out, low performing, etc). Though the average quality of reprocessed (recycled/remanufactured) goods (QO) is a criterion that can compare two or more candidate facilities, it is not justified to use QO as an independent criterion for comparison because QO depends on average quality of incoming products (QI). Thus, the idea is to take the difference between QO and QI as a criterion in the hierarchy.

- *Average supply of used products*: The only driver to design a forward supply chain is the demand for new products, so, if there is low demand for new products, there is practically no forward supply chain. However, this is not the case in some reverse supply chains where, even if there is low supply of used products (SU), a reverse supply chain must be administered due to the drivers, such as environmental regulations and asset recovery. In supply-driven cases like these, it is unfair to judge a recovery facility without considering SU in the hierarchy. Though throughput (TP) is a criterion that can compare two or more candidate recovery facilities, it is not justified to use TP as an independent criterion because TP depends on SU. In other words, a low SU might lead to a low TP, and a high SU might lead to a high TP. Thus, the idea is to take the ratio of TP to SU as a criterion in the hierarchy. The effect of a low TP is compensated for by dividing TP with a possibly low SU (by doing so, the facility under consideration is not underestimated). Similarly, the effect of a high TP is dampened by dividing TP with a possibly high SU (by doing so, the facility under consideration is not overestimated).

- *Average disassembly time of used products*: The average disassembly time (DT) is not exactly the inverse of TP because TP takes into account the whole reprocessing (disassembly plus recycling/remanufacturing) time. Unlike in a forward supply chain, components of incoming goods (used products) in a recovery facility are likely to be deformed or broken or different in number, even for the same type of products. Hence, incoming products of the same type might have different reprocessing times (unlike in a forward supply chain, where manufacturing time and assembly time are predetermined and equal for products of the same type). Because TP of a recovery facility depends upon DT, it is unfair not to consider DT in the hierarchy. In other words, a high DT might lead to a low TP, and a low DT might lead to a high TP. Thus, the idea is to use the multiplication of TP and DT as a criterion in the hierarchy. The effect of a low TP is compensated for by multiplying TP with a possibly high DT

(by doing so, the facility under consideration is not underestimated). Similarly, the effect of a high *TP* is dampened by multiplying *TP* with a possibly low *DT* (by doing so, the facility under consideration is not overestimated).

An intangible criterion considered in this model is termed customer service (*CS*). *CS* is a measure of how well a recovery facility utilizes the incentives provided by the government, to what extent it meets the environmental regulations, what kind of incentives it gives the collection centers that supply the used products, and what kind of incentives it is giving the customers buying the reprocessed goods. The all-encompassing term *customer service* is used here because any beneficiary is a customer, be it the government, the collection center, or the actual customer buying the reprocessed goods.

7.2.2   Numerical Example

Consider the following example: Table 7.1 shows the pair-wise comparison matrix for evaluation of candidate recovery facilities. It also shows the normalized eigen vector of the matrix. This vector represents the impacts given by the decision maker to the criteria.

Tables 7.2–7.6 show comparative ratings of the candidate recovery facilities, A, B, C, and D, with respect to the evaluation criteria *CO*, *QO–QI*, *TP/SU*, *TP×DT*, and *CS*, respectively. They also show the normalized eigen vectors representing the relative ratings with respect to the criteria. Note that each of the matrices in tables 7.1–7.6 has a consistency ratio *CR* of less than or equal to 0.1. For example, for the matrix in table 7.4, *CR*, using equation (4.1), is

$$\frac{(\lambda_{\max}-n)}{(n-1)(R)}=\frac{4.15-4}{(4-1)\times 0.90}=0.06.$$

Table 7.7 shows the aggregate matrix of relative ratings of recovery facilities with respect to each criterion in the second level of the hierarchy. This matrix is the collection of the eigen vectors obtained in tables 7.2–7.6.

TABLE 7.1

Comparative matrix for criteria (first model)

Criteria	*CO*	*QO–QI*	*TP/SU*	*TP×DT*	*CS*	Normalized eigen vector
CO	1	1/5	3	1	1/5	0.104
QO–QI	5	1	7	3	5	0.491
(*TP*)/(*SU*)	1/3	1/7	1	1/2	1/3	0.055
(*TP*)×(*DT*)	1	1/3	2	1	1/3	0.110
CS	5	1/5	3	3	1	0.240

TABLE 7.2
Comparative ratings of recovery facilities with respect to *CO* (first model)

CO Facilities	A	B	C	D	Normalized eigen vector
A	1	3	6	2	0.460
B	1/3	1	7	3	0.310
C	1/6	1/7	1	1/4	0.050
D	1/2	1/3	4	1	0.180

TABLE 7.3
Comparative ratings of recovery facilities with respect to *QO–QI* (first model)

QO–QI Facilities	A	B	C	D	Normalized eigen vector
A	1	1	7	4	0.380
B	1	1	7	7	0.445
C	1/7	1/7	1	1/5	0.050
D	1/4	1/7	5	1	0.125

TABLE 7.4
Comparative ratings of recovery facilities with respect to *TP/SU* (first model)

TP/SU Facilities	A	B	C	D	Normalized eigen vector
A	1	1/7	1/3	1/2	0.072
B	7	1	2	7	0.574
C	3	1/2	1	1	0.212
D	2	1/7	1	1	0.142

TABLE 7.5
Comparative ratings of recovery facilities with respect to *TP×DT* (first model)

TP×DT Facilities	A	B	C	D	Normalized eigen vector
A	1	1/5	1/2	1/2	0.091
B	5	1	3	7	0.595
C	2	1/3	1	1	0.171
D	2	1/7	1	1	0.143

TABLE 7.6

Comparative ratings of recovery facilities with respect to CS (first model)

Facilities	A	B	C	D	Normalized eigen vector
A	1	1/6	1/3	1/7	0.053
B	6	1	5	1/3	0.298
C	3	1/5	1	1/6	0.101
D	7	3	6	1	0.548

TABLE 7.7

Aggregate of ratings of recovery facilities (first model)

	A	B	C	D
CO	0.460	0.310	0.050	0.180
QO–QI	0.380	0.445	0.050	0.125
TP/SU	0.072	0.574	0.212	0.142
TP×DT	0.091	0.595	0.171	0.143
CS	0.053	0.298	0.101	0.548

By multiplying the matrix in table 7.7 with the normalized eigen vector obtained in table 7.1, the following normalized ranks are obtained for the recovery facilities: $\text{Rank}_A = 0.26$, $\text{Rank}_B = 0.42$, $\text{Rank}_C = 0.09$, and $\text{Rank}_D = 0.23$. If the decision maker desires to choose only those recovery facilities that have ranks of at least 0.25, he or she would choose recovery facilities A and B.

7.3   Second Model (Linear Physical Programming)

In this section, the model to identify efficient recovery facilities using linear physical programming (LPP) is presented. Section 7.3.1 presents the nomenclature for the LPP model, section 7.3.2 presents the evaluation criteria that are considered in this model, and section 7.3.3 presents a numerical example.

7.3.1   Nomenclature for LPP Model

CS_v Customer service rating of recovery facility v (numerical scale)

DT_v Average disassembly time of products supplied to recovery facility v (time units)

g_i ith criterion for evaluation of candidate recovery facilities

IT_{uv} Transit time between collection center u and recovery facility v (time units)

K Transportation cost per unit time ($ per unit time)

L_v Labor cost at location of recovery facility v ($ per unit time)

M_v Inventory (space) cost at recovery facility v ($ per unit area)

OT_{vw} Transit time between recovery facility v and demand center w (time units)

QI_v Average quality of products supplied to recovery facility v (numerical scale)

QO_v Average quality of outgoing products from recovery facility v (numerical scale)

R_v Reprocessing cost at recovery facility v ($ per unit product)

SU_v Supply to recovery facility v (products)

TP_v Throughput of recovery facility v (products)

u Collection center

v Recovery facility

w Demand center

7.3.2 Criteria for Identification of Efficient Recovery Facilities

7.3.2.1 Class 1S Criteria (Smaller is Better)

The cost of transporting goods (used as well as reprocessed), g_1, through a recovery facility v across the reverse supply chain is calculated using the following equation:

$$g_1 = \sum_u (IT_{uv})(K) + \sum_w (OT_{vw})(K) \qquad (7.1)$$

The operating cost, g_2, incurred by a recovery facility v is the sum of the labor cost, the inventory cost, and the reprocessing cost. Thus,

$$g_2 = L_v + M_v + R_v \qquad (7.2)$$

7.3.2.2 Class 2S Criteria (Larger Is Better)

As explained in section 7.2, the difference between QO and QI is taken as a criterion for evaluation. Thus,

$$g_3 = QO_v - QI_v \qquad (7.3)$$

Also, *TP* is divided by *SU* and then taken as a criterion for evaluation:

$$g_4 = TP_v / SU_v \tag{7.4}$$

Similarly,

$$g_5 = TP_v \times DT_v \tag{7.5}$$

Customer service (*CS*) is taken as a criterion for evaluation in the physical programming model as well. Thus,

$$g_6 = CS_v \tag{7.6}$$

7.3.3 Numerical Example

Three candidate recovery facilities (E, F, and G) are evaluated using the LPP model and then ranked to identify the efficient ones.

Table 7.8 shows the target values for each criterion detailed in section 7.3.2. Table 7.9 shows the criteria values for each recovery facility. Table 7.10 shows the incremental weights obtained by using the LPP weight algorithm [2]. Tables 7.11–7.13 show the deviations of criteria values from the target values for facilities E, F, and G, respectively. Table 7.14 shows the total scores (obtained using equation (4.34)) and the ranks of the recovery facilities. It is obvious from table 7.14 that G is the most desirable facility and F is the least desirable. If the decision maker has a cutoff limit of, say, 100, he or she will identify facilities E and G as efficient.

7.4   Evaluation Criteria for Third and Fourth Models

Like in evaluation of collection centers (see chapter 6), in evaluation of recovery facilities one may have a scenario with three different categories of decision makers with multiple, conflicting, and incommensurate goals, as follows:

TABLE 7.8

Preference table (second model)

Criteria	$t_{p1}+$	$t_{p2}+$	$t_{p3}+$	$t_{p4}+$	$t_{p5}+$
g_1	10	15	25	30	45
g_2	12	14	15	16	17
Criteria	$t_{p1}-$	$t_{p2}-$	$t_{p3}-$	$t_{p4}-$	$t_{p5}-$
g_3	0.6	0.4	0.3	0.2	0
g_4	1.1	0.9	0.7	0.6	0.4
g_5	250	200	140	120	100
g_6	10	7	6	4	3

TABLE 7.9

Criteria values for each recovery facility (second model)

Criteria	Facility E	Facility F	Facility G
g_1	22	30	15
g_2	15	17	10
g_3	0.4	0.3	0.1
g_4	0.5	0.8	0.5
g_5	200	220	145
g_6	8	6	4

TABLE 7.10

Output of LPP weight algorithm (second model)

Criteria	$\Delta w_{p2}+$	$\Delta w_{p3}+$	$\Delta w_{p4}+$	$\Delta w_{p5}+$	$\Delta w_{p2}-$	$\Delta w_{p3}-$	$\Delta w_{p4}-$	$\Delta w_{p5}-$
g_1	0.02	0.012	0.168	0.011	—	—	—	—
g_2	0.05	0.266	0.683	2.157	—	—	—	—
g_3	—	—	—	—	0.05	0.115	0.380	1.252
g_4	—	—	—	—	0.05	0.280	0.215	1.252
g_5	—	—	—	—	0.033	0.077	0.253	0.835
g_6	—	—	—	—	0.033	0.077	0.979	2.505

TABLE 7.11

Deviations of criteria values of recovery facility E from target values (second model)

Criteria	$r = 2$	$r = 3$	$r = 4$	$r = 5$
g_1	$d_{12}+ = 12$	$d_{13}+ = 7$	$d_{14}+ = 3$	$d_{15}+ = 8$
g_2	$d_{22}- = 3$	$d_{23}- = 1$	$d_{24}- = 0$	$d_{25}- = 1$
g_3	$d_{32}- = 0.2$	$d_{33}- = 0$	$d_{34}- = 0.1$	$d_{35}- = 0.2$
g_4	$d_{42}- = 0.6$	$d_{43}- = 0.4$	$d_{44}- = 0.2$	$d_{45}- = 0.1$
g_5	$d_{52}- = 50$	$d_{53}- = 0$	$d_{54}- = 60$	$d_{55}- = 80$
g_6	$d_{62}- = 2$	$d_{63}- = 1$	$d_{64}- = 2$	$d_{65}- = 4$

TABLE 7.12

Deviations of criteria values of recovery facility F from target values (second model)

Criteria	$r = 2$	$r = 3$	$r = 4$	$r = 5$
g_1	$d_{12}+ = 20$	$d_{13}+ = 15$	$d_{14}+ = 5$	$d_{15}+ = 0$
g_2	$d_{22}- = 0.3$	$d_{23}- = 0.1$	$d_{24}- = 0$	$d_{25}- = 0.1$
g_3	$d_{32}- = 0.3$	$d_{33}- = 0.1$	$d_{34}- = 0.1$	$d_{35}- = 0.2$
g_4	$d_{42}- = 30$	$d_{43}- = 20$	$d_{44}- = 80$	$d_{45}- = 100$
g_5	$d_{52}- = 4$	$d_{53}- = 1$	$d_{54}- = 0$	$d_{55}- = 2$

TABLE 7.13

Deviations of criteria values of recovery facility G from target values (second model)

Criteria	$r = 2$	$r = 3$	$r = 4$	$r = 5$
g_1	$d_{12}+ = 5$	$d_{13}+ = 0$	$d_{14}+ = 10$	$d_{15}+ = 15$
g_2	$d_{22}- = 0.5$	$d_{23}- = 0.3$	$d_{24}- = 0.2$	$d_{25}- = 0.1$
g_3	$d_{32}- = 0.6$	$d_{33}- = 0.4$	$d_{34}- = 0.2$	$d_{35}- = 0.1$
g_4	$d_{42}- = 105$	$d_{43}- = 55$	$d_{44}- = 5$	$d_{45}- = 25$
g_5	$d_{52}- = 6$	$d_{53}- = 3$	$d_{54}- = 2$	$d_{55}- = 0$

TABLE 7.14

Total scores and ranks of recovery facilities (second model)

Recovery facilities	Total scores	Ranks
E	99.85	II
F	117.95	III
G	52.26	I

- *Consumers* whose primary concern is *convenience*
- *Local government officials* whose primary concern is *environmental consciousness*
- *Supply chain company executives* whose primary concern is *profit*

Therefore, the efficiency of a candidate recovery facility must be evaluated based on the maximized consensus among decision makers of the three categories. Sections 7.4.1, 7.4.2, and 7.4.3 present the lists of criteria that are considered for the three categories.

7.4.1   Criteria of Consumers

- Proximity to surface water (*PS*) (the closer the facility is to the surface water, the higher the facility's suitability, i.e., less hazardous)
- Proximity to residential area (*PH*) (the closer the facility is to the residential area, the higher the facility's suitability, i.e., less hazardous)
- Employment opportunity (*EO*) (the higher, the better)
- Salary offered to employees at recovery facility (*SA*) (the higher, the better)

7.4.2   Criteria of Local Government Officials

- Proximity to surface water (*PS*) (the closer the facility is to surface water, the higher the facility's suitability, i.e., less hazardous)

- Proximity to residential area (*PH*) (the closer the facility is to the residential area, the higher the facility's suitability, i.e., less hazardous)

7.4.3   Criteria of Supply Chain Company Executives

- Space cost (*SC*) (the lower, the better)
- Labor cost (*LC*) (the lower, the better)
- Proximity to roads (*PR*) (the closer the facility is to roads, the easier is the transportation)
- Quality of reprocessed products (*QO*)–quality of used products (*QI*) (the higher, the better)
- Throughput (*TP*)/supply (*SU*) (the higher, the better)
- Throughput (*TP*)×disassembly time (*DT*) (the higher, the better)
- Utilization of incentives from local government (*UI*) (the higher, the better)
- Pollution control (*PC*) (the higher, the better)

7.5   Third Model (Eigen Vector Method, TOPSIS, and Borda's Choice Rule)

This model to select efficient recovery facilities is implemented in two phases. In the first phase, using the eigen vector method, impacts are given to the criteria identified for each category of decision makers (see section 7.4), and then TOPSIS (technique for order preference by similarity to ideal solution) is employed to find the efficiency of each candidate recovery facility, as evaluated by that category. In the second phase, Borda's choice rule is used to combine individual evaluations for each candidate recovery facility into a group evaluation or maximized consensus ranking.

The model to select efficient recovery facilities is presented using a numerical example. Three recovery facilities, H, J, and K, are considered for evaluation.

7.5.1   Phase I (Individual Decision Making)

Tables 7.15–7.17 show the pair-wise comparison matrices as formed for the consumers, local government officials, and supply chain company executives, respectively.

Tables 7.18–7.20 show the impacts given for the criteria of the consumers, local government officials, and supply chain company executives, respectively. These sets of impacts, calculated using the eigen vector method, are the elements of the normalized eigen vectors of pair-wise comparison matrices shown in tables 7.15–7.17, respectively. For example, the relative impacts

TABLE 7.15

Pair-wise comparison matrix for consumers (third model)

Criteria	PS	PH	EO	SA
PS	1	1/2	1/2	1/3
PH	2	1	1	1
EO	2	1	1	1
SA	3	1	1	1

TABLE 7.16

Pair-wise comparison matrix for local
government officials (third model)

Criteria	PS	PH
PS	1	1/3
PH	3	1

TABLE 7.17

Pair-wise comparison matrix for supply chain company executives (third model)

Criteria	SC	LC	PR	QO–QI	TP/SU	TP×DT	UI	PC
SC	1	1	1	1	½	6	1	1
LC	1	1	1	1	1	5	2	1
PR	1	1	1	1/9	1/7	1	1	1/3
QO–QI	1	1	1	1	2	2	1	1/4
TP/SU	2	1	7	1	1	5	1	1
TP×DT	1/6	1/5	1	1/2	0.2	1	1/9	1
UI	1	1/2	1	1	1	9	1	1
PC	1	1	3	4	1	1	1	1

of the criteria *PH* (proximity to residential area) and *PR* (proximity to roads) shown in table 7.19 are the elements of the normalized eigen vector of the pair-wise comparison matrix shown in table 7.16.

The decision matrices formed for the consumers, local government officials, and supply chain company executives, for implementation of the TOPSIS for each category, are shown in tables 7.21–7.23, respectively. The elements of these matrices are the ranks (ranging from 1 to 10) assigned to the recovery facilities with respect to each criterion for evaluation. A lower rank implies higher efficiency (with respect to that criterion).

To facilitate the construction of the pair-wise comparison matrices (see tables 7.15–7.17) and the decision matrices (see tables 7.21–7.23), representatives from each category of decision makers could be invited to participate in relevant survey questionnaires, individual interviews, focus groups, and on-site observations.

TABLE 7.18

Impacts for consumers (third model)

Criteria	Impacts
PS	0.1277
PH	0.2804
EO	0.2804
SA	0.3116

TABLE 7.19

Impacts for local government officials
(third model)

Criteria	Impacts
PS	0.25
PH	0.75

TABLE 7.20

Impacts for supply chain company
executives (third model)

Criteria	Impacts
SC	0.1233
LC	0.1437
PR	0.0717
QO–QI	0.1198
TP/SU	0.1905
TP×DT	0.0491
UI	0.1356
PC	0.1663

Now one is ready to perform the six steps in the TOPSIS for each category of decision makers. The following steps show the implementation of the TOPSIS for the consumers to evaluate H, J, and K.

Step 1: *Construct the normalized decision matrix.* Table 7.24 shows the normalized decision matrix formed by applying equation (4.49) on each element of table 7.21 (decision matrix formed for the consumers). For example, the normalized rank of recovery facility J with respect to criterion *PH* (see tables 7.21 and 7.24) is calculated as follows:

$$r_{22} = \frac{1}{\sqrt{2^2 + 1^2 + 3^2}} = 0.2673.$$

TABLE 7.21

Decision matrix for consumers (third model)

Recovery facilities	PS	PH	EO	SA
H	2	2	4	1
J	4	1	7	3
K	5	3	3	1

TABLE 7.22

Decision matrix for local government officials (third model)

Recovery facilities	PS	PH
H	4	2
J	3	1
K	1	5

TABLE 7.23

Decision matrix for supply chain company executives (third model)

Recovery facilities	SC	LC	PR	QO–QI	TP/SU	TP×DT	UI	PC
H	3	2	1	3	4	1	1	2
J	4	7	1	2	2	2	3	4
K	9	8	7	5	1	6	5	6

TABLE 7.24

Normalized decision matrix for consumers (third model)

Recovery facilities	PS	PH	EO	SA
H	0.2981	0.5345	0.4650	0.3015
J	0.5963	0.2673	0.8137	0.9045
K	0.7454	0.8018	0.3487	0.3015

Step 2: *Construct the weighted normalized decision matrix.* Table 7.25 shows the weighted normalized decision matrix for the consumers. This is constructed using the impacts of the criteria listed in table 7.18 and the normalized decision matrix in table 7.24. For example, the weighted normalized rank of recovery facility J with respect to criterion *PH*, i.e., 0.0749 (see table 7.25), is calculated by multiplying the impact of *PH*, i.e., 0.2804 (see table 7.18), by the normalized rank of R2 with respect to *PH*, i.e., 0.2673 (see table 7.24).

TABLE 7.25

Weighted normalized decision matrix for consumers (third model)

Recovery facilities	PS	PH	EO	SA
H	0.0381	0.1499	0.1304	0.0940
J	0.0761	0.0749	0.2282	0.2819
K	0.0952	0.2248	0.0978	0.0940

TABLE 7.26

Separation distances for consumers (third model)

Recovery facilities	S*	S–
H	0.0817	0.2318
J	0.2318	0.1511
K	0.1604	0.2287

TABLE 7.27

Relative closeness coefficients for consumers (third model)

Recovery facilities	C*
H	0.7394
J	0.3945
K	0.5878

Step 3: *Determine the ideal and negative-ideal solutions.* Each column in the weighted normalized decision matrix shown in table 7.25 has a minimum rank and a maximum rank. They are the ideal and negative-ideal solutions, respectively, for the corresponding criterion. For example (see table 7.25), with respect to criterion *PH*, the ideal solution (minimum rank) is 0.0749, and the negative-ideal solution (maximum rank) is 0.2248.

Step 4: *Calculate the separation distances.* The separation distances (see table 7.26) for each recovery facility are calculated using equations (4.52) and (4.53). For example, the positive separation distance for recovery facility J (see table 7.26) is calculated using equation (4.52), which contains the weighted normalized ranks of J (see table 7.25) and the ideal solutions (obtained in step 3) for the criteria.

Step 5: *Calculate the relative closeness coefficient.* Using equation (4.54), the relative closeness coefficient is calculated for each recovery facility (see table 7.27). For example, the relative closeness coefficient (i.e., 0.3945) for recovery facility J (see table 7.27) is the ratio of J's negative separation distance (i.e., 0.1511) to the sum (i.e., 0.1511 + 0.2318 = 0.3829) of its negative and positive separation distances (see table 7.26).

Step 6: *Form the preference order.* Because the best alternative is the one with the highest relative closeness coefficient, the preference order for the recovery facilities is H, K, and J (that means H is the best recovery facility, as evaluated by the consumers).

TOPSIS is implemented for the local government officials and supply chain company executives in a similar manner. The relative closeness coefficients of the recovery facilities, as calculated for those two categories of decision makers, are shown in table 7.28.

7.5.2   Phase II (Group Decision Making)

Table 7.29 shows the marks of the recovery facilities as given using Borda's choice rule for the consumers, local government officials, and supply chain company executives. Borda scores (group evaluations) calculated for H, J, and K (viz., 5, 4, and 2, respectively) are also shown. For example, the Borda score for J (i.e., 4) is calculated by summing the marks of J for the consumers, local government officials, and supply chain company executives (i.e., 0 + 2 + 2). Because H has the highest Borda score, it is the best of the lot.

7.6   Fourth Model (Neural Networks, Fuzzy Logic, TOPSIS, Borda's Choice Rule)

This model assumes that neither numerical nor linguistic impacts are available for the evaluation criteria. It employs a neural network [6] to evaluate

TABLE 7.28

Relative closeness coefficients for local government officials and supply chain company executives (third model)

Recovery facilities	Local government officials	Supply chain company executives
H	0.6715	0.5952
J	0.8487	0.5962
K	0.2941	0.4057

TABLE 7.29

Marks and Borda scores of recovery facilities (third model)

Recovery facilities	Consumers	Local government officials	Supply chain company executives	Borda scores
H	2	1	2	5
J	0	2	2	4
K	1	0	1	2

the efficiency of a recovery facility of interest (which is being considered for inclusion in a reverse supply chain), using linguistic performance measures of recovery facilities that already exist in the reverse supply chain. To this end, the model to evaluate the efficiency of a recovery facility of interest is carried out in three phases. In the first phase, the ratings of existing recovery facilities are used to construct a neural network that, in turn, calculates impacts of criteria identified for each category of decision makers given in section 7.4. Then, in the second phase, the impacts obtained in the first phase are used in a fuzzy TOPSIS (combination of fuzzy logic and TOPSIS) method to obtain the overall rating of the recovery facility of interest, as calculated for each category. Finally, in the third phase, Borda's choice rule is employed to calculate the maximized consensus rating (among the categories considered), i.e., efficiency, of the recovery facility of interest.

7.6.1 Phase I (Derivation of Impacts)

Suppose that one has the linguistic ratings of ten existing recovery facilities, as given by an expert in each category of decision makers described in section 7.4. Using fuzzy logic, these linguistic ratings are converted into triangular fuzzy numbers (TFNs). Table 7.30 shows not only one of the many ways for conversion of linguistic ratings into TFNs but also the defuzzified ratings of the corresponding TFNs. Tables 7.31–7.33 show the defuzzified overall rating of each existing recovery facility as well as the recovery facility's defuzzified rating with respect to each criterion, as evaluated by the consumers, local government officials, and supply chain company executives, respectively. Defuzzification of a TFN can be performed using equation (4.8).

A neural network is constructed and trained for each category of decision makers, using the defuzzified ratings of the existing recovery facilities with respect to criteria as input sets and the recovery facilities' defuzzified overall ratings as corresponding outputs. In the example, there are ten input–output pairs for each neural network because there are ten existing recovery facilities. Also, three layers are considered in each network, with five nodes in the hidden layer. The number of nodes in the output layer is one (for overall rating), and the number in the input layer is the number of criteria

TABLE 7.30

Conversion table for ratings (fourth model)

Linguistic ratings	TFNs	Defuzzified ratings
Very good (VG)	(7, 10, 10)	9
Good (G)	(5, 7, 10)	7.3
Fair (F)	(2, 5, 8)	5
Poor (P)	(1, 3, 5)	3
Very poor (VP)	(0, 0, 3)	1

TABLE 7.31

Consumer ratings of recovery facilities (fourth model)

Recovery facilities	PS	PH	EO	SA	Overall
R1	9	3	7.33	3	3
R2	5	7.33	9	1	1
R3	1	5	1	5	9
R4	1	5	5	9	5
R5	5	9	5	3	7.33
R6	1	3	5	3	5
R7	9	1	3	5	1
R8	3	1	3	1	9
R9	3	9	1	7.33	5
R10	5	1	3	5	7.33

TABLE 7.32

Local government officials' ratings of recovery facilities (fourth model)

Recovery facilities	PS	PH	Overall
R1	9	3	5
R2	5	7.33	3
R3	1	5	9
R4	1	5	7.33
R5	5	9	1
R6	1	3	1
R7	9	1	7.33
R8	3	1	5
R9	3	9	5
R10	5	1	3

TABLE 7.33

Supply chain company executives' ratings of recovery facilities (fourth model)

Recovery facilities	SC	LC	PR	QO–QI	TP/SU	TP×DT	UI	PC	Overall
R1	9	3	9	5	3	9	3	3	5
R2	5	7.33	3	1	7.33	9	1	1	3
R3	1	5	1	3	1	3	5	9	1
R4	1	5	3	5	9	9	1	5	7.33
R5	5	9	1	3	5	7.33	7.33	1	9
R6	9	3	9	5	3	9	3	3	9
R7	5	7.33	3	1	7.33	9	1	1	1
R8	1	5	1	3	1	3	5	9	3
R9	1	5	3	5	9	9	1	5	5
R10	5	9	1	3	5	7.33	7.33	1	7.33

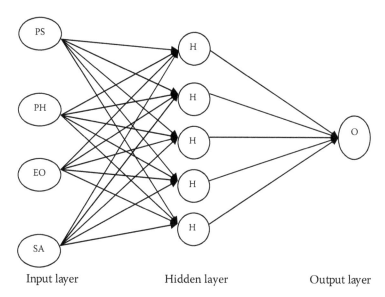

FIGURE 7.2
Neural network for consumers (fourth model).

considered by the corresponding category. For example, figure 7.2 shows the neural network constructed and trained for the consumer category.

After each neural network is trained, the following equation [1] is used to calculate the impacts of criteria considered by the corresponding category. Here, the absolute value of W_v is the impact of the vth input node upon the output node, n_V is the number of input nodes, n_H is the number of hidden nodes, I_{ij} is the connection weight from the ith input node to the jth hidden node, and O_j is the connection weight from the jth hidden node to the output node:

$$|W_v| = \frac{\sum_{j}^{n_H} \frac{I_{vj}}{\sum_{i}^{n_V} |I_{ij}|} O_j}{\sum_{i}^{n_V} \left(\sum_{j}^{n_H} \left(\frac{I_{vj}}{\sum_{i}^{n_V} |I_{ij}|} O_j \right) \right)}$$

(7.7)

Tables 7.34–7.36 show the impacts of the criteria considered for the consumers, local government officials, and supply chain company executives, respectively.

TABLE 7.34

Impacts of criteria of consumers (fourth model)

Criteria	PS	PH	EO	SA
Impacts	0.17	0.18	0.27	0.38

TABLE 7.35

Impacts of criteria of local government officials (fourth model)

Criteria	PS	PH
Impacts	0.61	0.39

TABLE 7.36

Impacts of criteria of supply chain company executives (fourth model)

Criteria	SC	LC	PR	QO–QI	TP/SU	TP×DT	UI	PC
Impacts	0.06	0.02	0.10	0.19	0.11	0.39	0.10	0.02

TABLE 7.37

Decision matrix for consumers (fourth model)

Recovery facilities	PS	PH	EO	SA
R11	1	5	1	5
R12	1	5	5	9
R13	5	9	5	3

TABLE 7.38

Decision matrix for local government officials (fourth model)

Recovery facilities	PS	PH
R11	3	1
R12	3	9
R13	5	1

TABLE 7.39

Decision matrix for supply chain company executives (fourth model)

Recovery facilities	SC	LC	PR	QO–QI	TP/SU	TP×DT	UI	PC
R11	5	9	1	3	5	7.33	7.33	1
R12	9	3	9	5	3	9	3	3
R13	5	7.33	3	1	7.33	9	1	1

7.6.2 Phase II (Individual Decision Making)

Suppose that there are three recovery facilities, R11, R12, and R13, of interest. A fuzzy TOPSIS method uses the impacts obtained in the first phase to calculate the overall ratings of the three recovery facilities.

The decision matrices formed for the consumers, local government officials, and supply chain company executives (with defuzzified ratings for R11, R12, and R13) in this example are shown in tables 7.37–7.39, respectively (table 7.30 is used here as well, to convert linguistic ratings given by each category into TFNs). Like in the third model (see section 7.5), the matrices can be constructed by inviting representatives from each category of decision makers to participate in relevant survey questionnaires, individual interviews, focus groups, and on-site observations.

Now one is ready to perform the six steps in the TOPSIS for each category of decision makers. The following steps show the implementation of the TOPSIS for the consumers to evaluate R11, R12, and R13.

Step 1: *Construct the normalized decision matrix.* Table 7.40 shows the normalized decision matrix formed by applying equation (4.49) on each element of table 7.37 (decision matrix for the consumers). For example, the normalized rating of recovery facility R12 with respect to criterion *PH* (see tables 7.37 and 7.40) is calculated as follows:

$$r_{22} = \frac{5}{\sqrt{5^2 + 5^2 + 9^2}} = 0.4369.$$

Step 2: *Construct the weighted normalized decision matrix.* Table 7.41 shows the weighted normalized decision matrix for the consumers. This is constructed using the impacts of the criteria listed in table 7.34 and the normalized decision matrix in table 7.40. For example, the weighted normalized rank of recovery facility R12 with respect to criterion *PH*, i.e., 0.0793 (see table 7.41), is calculated by multiplying the impact of *PH*, i.e., 0.18 (see table 7.34), with the normalized rating of R12 with respect to *PH*, i.e., 0.4369 (see table 7.40).

Step 3: *Determine the ideal and negative-ideal solutions.* Each column in the weighted normalized decision matrix shown in table 7.41 has a maximum rating and a minimum rating. They are the ideal and negative-ideal solutions, respectively, for the corresponding criterion. For example (see table 7.41), with respect to criterion *PH*, the ideal solution (maximum rating) is 0.1428, and the negative-ideal solution (minimum rating) is 0.0793.

Step 4: *Calculate the separation distances.* The separation distances (see table 7.42) for each recovery facility are calculated using equations (4.52) and (4.53). For example, the positive separation distance for

TABLE 7.40

Normalized decision matrix for consumers (fourth model)

Recovery facilities	PS	PH	EO	SA
R11	0.1923	0.4369	0.1400	0.4663
R12	0.1925	0.4369	0.7001	0.8393
R13	0.9623	0.7863	0.7001	0.2798

TABLE 7.41

Weighted normalized decision matrix for consumers (fourth model)

Recovery facilities	PS	PH	EO	SA
R11	0.0330	0.0793	0.0371	0.1780
R12	0.0330	0.0793	0.1857	0.3203
R13	0.1650	0.1428	0.1857	0.1068

TABLE 7.42

Separation distances for consumers (fourth model)

Recovery facilities	S*	S−
R11	0.2526	0.0712
R12	0.1464	0.2602
R13	0.2136	0.2086

recovery facility R12 (see table 7.42) is calculated using equation (4.52), which contains the weighted normalized ratings of R12 (see table 7.41) and the ideal solutions (obtained in step 3) for the criteria.

Step 5: *Calculate the relative closeness coefficient.* Using equation (4.54), the relative closeness coefficient is calculated for each recovery facility (see table 7.43). For example, the relative closeness coefficient (i.e., 0.6398) for recovery facility R12 (see table 7.43) is the ratio of R12's negative separation distance (i.e., 0.2602) to the sum (i.e., 0.2602 + 0.1464 = 0.4066) of its negative and positive separation distances (see table 7.42).

Step 6: *Rank the preference order.* Because the best alternative is the one with the highest relative closeness coefficient, the preference order for the recovery facilities is R12, R13, and R11 (that means R12 is the best recovery facility, as evaluated by the consumers).

Similarly, TOPSIS is implemented for the local government officials and supply chain company executives. The relative closeness coefficients of the recovery facilities, as calculated for those two categories, are shown in table 7.44.

TABLE 7.43

Relative closeness coefficients for consumers (fourth model)

Recovery facilities	C^*
R11	0.2199
R12	0.6398
R13	0.4942

TABLE 7.44

Relative closeness coefficients for local government officials and supply chain company executives (fourth model)

Recovery facilities	Local government officials	Supply chain company executives
R11	0	0.458016
R12	0.646968	0.695261
R13	0.588066	0.29753

7.6.3 Phase III (Group Decision Making)

Table 7.45 shows the marks of the recovery facilities as given using Borda's choice rule for the consumers, local government officials, and supply chain company executives. Borda scores (group evaluations) calculated for R11, R12, and R13 (viz., 1, 6, and 2, respectively) are also shown. For example, the Borda score for R12 (i.e., 6) is calculated by summing the marks of R12 for the consumers, local government officials, and supply chain company executives (i.e., 2 + 2 + 2). Because R12 has the highest Borda score, it is the best of the lot.

TABLE 7.45

Marks and Borda scores of recovery facilities (fourth model)

Recovery facilities	Consumers	Local government officials	Supply chain company executives	Borda scores
R11	0	0	1	1
R12	2	2	2	6
R13	1	1	0	2

7.7   Fifth Model (Two-Dimensional Chart)

Here, a simple two-dimensional chart, where only the two most important evaluation criteria are considered, is presented. In this approach, it is assumed that the two most important criteria are (1) n value in n Sigma of the recovery facility and (2) TP/SU value.

This approach is similar (but with some important modifications) to Linn et al.'s approach [5] for selection of suppliers in a traditional supply chain. Linn et al.'s approach uses C_{pk} (process capability index) and a price comparison (CPC) chart for the selection of efficient suppliers (see section 4.16 for the concept of process capability). The CPC chart (see figure 7.3), which integrates the process capability and price information of multiple suppliers (assuming every other criterion value is either unimportant or the same for all suppliers), provides a method to consider quality and price simultaneously in the supplier selection process.

C_{pk} is drawn on the Y-axis, and the ratio (R) of target price desired to price quoted by the supplier is drawn on the X-axis. C_{pk} and R of each supplier are then plotted on the chart. The chart is partitioned into six different zones representing the quality performance and price levels. The zones are defined as follows:

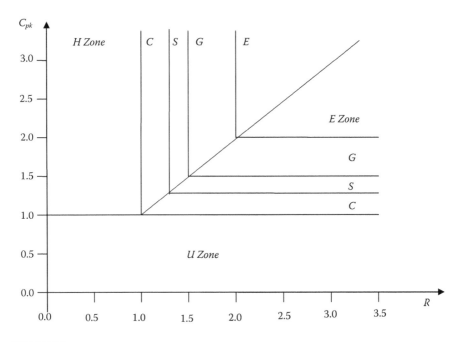

FIGURE 7.3
CPC chart

E zone: Excellent zone, $E = \{(R, C_{pk}) \mid C_{pk} > 2.0 \text{ and } R > 2.0\}$

G zone: Good zone, $G = \{(R, C_{pk}) \mid C_{pk} > 1.5 \text{ and } R > 1.5\} - E$ zone

S zone: Satisfactory zone, $S = \{(R, C_{pk}) \mid C_{pk} > 1.33 \text{ and } R > 1.33\} - G$ zone $- E$ zone

C zone: Capable zone, $C = \{(R, C_{pk}) \mid C_{pk} > 1.0 \text{ and } R > 1.0\} - G$ zone $- E$ zone $- S$ zone

H zone: High-price zone, $H = \{(R, C_{pk}) \mid C_{pk} > 1.0 \text{ and } R < 1.0\}$

U zone: Unacceptable zone, $U = \{(R, C_{pk}) \mid C_{pk} < 1.0 \text{ and } R > 0\}$

Because of their high-quality performance and low-cost quotation, those suppliers falling in the E zone are considered the best group of suppliers to choose from. Those falling in the U zone are simply not acceptable because their C_{pk} value is too low. The supplier selection should start from the E zone and follow the sequence of $E \rightarrow G \rightarrow S \rightarrow C \rightarrow H \rightarrow U$. Within the same zone, those falling in the lower-half zone (below the 45° line) have better cost performance. In contrast, those in the upper-half zone (above the 45° line) have better quality performance. Therefore, if the objective is to select a better-cost performer, those in the lower half of the zone should be selected. If the objective is to find a better-quality performer, those in the upper-half zone should be selected. If quality and cost are equally important, those close to the 45° line should be selected.

Although the above approach can integrate both quality and price, it must be noted that the C_{pk} value is not enough to judge quality; one needs the C_p (process capability ratio) value as well. To overcome this problem in the selection of efficient recovery facilities, the n value (instead of C_{pk}) is used on the Y-axis of the chart (see section 4.16 for the concepts of process capability and n Sigma).

The n value for a recovery facility represents the facility's quality, which could be a function of factors such as efficiency of reprocessing and efficiency of delivery (of reprocessed goods to demand centers). For example (see figure 7.4), let the upper specification (U) of the width (critical dimension) of a reprocessed good at a candidate recovery facility be 100 inches, the lower specification (L) 60 inches, and the target value (τ) 80 inches. If the standard deviation (σ) of the width is 5 inches and the mean shift ($|\tau - \mu|$) is 6 inches, then reprocessing in the candidate recovery facility is a Four Sigma process.

For the chart's X-axis, the TP/SU criterion is used. See figure 7.5 for the chart that is used for selection of efficient recovery facilities.

The chart is partitioned into five acceptable regions (I–V) and an unacceptable region. These zones are defined as follows:

Zone I $= \{(TP/SU, n) \mid 0.9 \leq TP/SU \leq 1 \text{ and } n \geq 5.0\}$

Zone II $= \{(TP/SU, n) \mid TP/SU \geq 0.8 \text{ and } n \geq 4.0\} -$ zone I

Zone III $= \{(TP/SU, n) \mid TP/SU \geq 0.7 \text{ and } n \geq 3.0\} -$ zone I $-$ zone II

Zone IV = {(TP/SU, n) | TP/SU ≥ 0.6 and n ≥ 2.0} – zone I – zone II – zone III

Zone V = {(TP/SU, n) | TP/SU ≥ 0.5 and n ≥ 2.0} – zone I – zone II – zone III – zone IV

Unacceptable zone = {(TP/SU, n) | TP/SU ≤ 0.5 or n < 2.0}

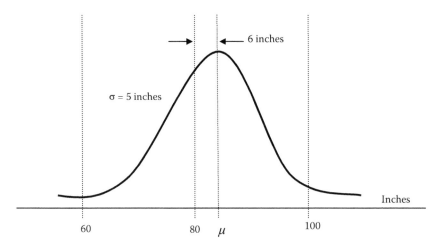

FIGURE 7.4
Critical dimension of a reprocessed good at recovery facility (fifth model).

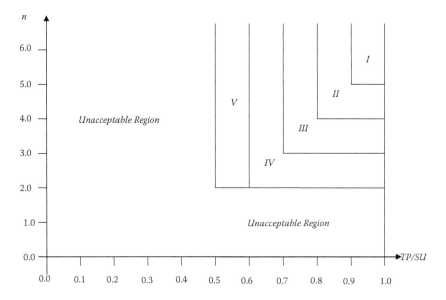

FIGURE 7.5
n and *TP/SU* chart (fifth model).

Because of their high-quality (Five Sigma or higher) performance and high *TP/SU* value, those recovery facilities falling in zone I are considered the best group of recovery facilities to choose from. Those falling in the unacceptable zone are simply not acceptable because their *n* value (2 or lower) is too low, their *TP/SU* value is low, or both. The recovery facility selection should start from zone I and follow the sequence of I → II → III → IV → V.

If the decision maker wishes to raise or lower the bar for quality, he or she can revise the respective charts accordingly. For example, if the decision maker considers only those recovery facilities with at least Six Sigma quality (besides high *TP/SU* ratio) the best, then zone I in his or her chart will be $\{(n, TP/SU) \mid n \geq 6.0 \text{ and } 0.9 \leq TP/SU \leq 1\}$.

7.8 Conclusions

In this chapter, five models are presented for identifying efficient recovery facilities in a region where a reverse supply chain is to be designed. The first model uses analytic hierarchy process in a situation where the evaluation criteria can be given numerical impacts. The second model employs linear physical programming in a situation where the evaluation criteria are presented in terms of ranges of different degrees of desirability. The third and fourth models have decision makers with conflicting evaluation criteria. Whereas the third model uses the eigen vector method, TOPSIS, and Borda's choice rule, the fourth model uses neural network, fuzzy logic, TOPSIS, and Borda's choice rule. Finally, the fifth model presents a two-dimensional chart that can be used when a very simple evaluation technique is desired (only the two most important evaluation criteria are considered).

References

1. Cha, Y., and Jung, M. 2003. Satisfaction assessment of multi-objective schedules using neural fuzzy methodology. *International Journal of Production Research* 41:1831–49.
2. Messac, A., Gupta, S. M., and Akbulut, B. 1996. Linear physical programming: A new approach to multiple objective optimization. *Transactions on Operational Research* 8:39–59.
3. Talluri, S., and Baker, R. C. 2002. A multi-phase mathematical programming approach for effective supply chain design. *European Journal of Operational Research* 141:544–58.
4. Talluri, S., Baker, R. C., and Sarkis, J. 1999. Framework for designing efficient value chain networks. *International Journal of Production Economics* 62:133–44.

5. Linn, R. J., Tsung, F., and Ellis, L. W. C. 2006. Supplier selection based on process capability and price analysis. *Quality Engineering* 18:123–29.
6. Haykin, S. 1998. *Neural networks: A comprehensive foundation*. 2nd ed. Englewood Cliffs, NJ: Prentice Hall.

8

Optimization of Transportation of Products

8.1 The Issue

In this chapter, the focus is on achieving transportation of the right quantities of products (used, remanufactured, and new) across a reverse or closed-loop supply chain while satisfying certain constraints. The various scenarios for this problem could differ as follows:

1. Decision-making criteria for a reverse supply chain could be given in terms of classical supply-and-demand constraints.
2. Decision-making criteria for a reverse supply chain could be presented in terms of ranges of different degrees of desirability.
3. Besides optimal transportation of products, there could be a need to address the following issues in one continuous phase for a closed-loop supply chain: selection of used products and evaluation of production facilities. Also, the decision-making criteria could be presented in terms of classic supply-and-demand constraints.
4. The situation could be the same as in scenario 3, but the decision-making criteria could be presented in terms of different degrees of desirability.
5. The situation could be the same as in scenario 3, but the decision-making criteria could be imprecise.

In this chapter, various models to address the above scenarios are presented. The first model addresses scenario 1 and employs linear integer programming. The second model is for scenario 2 and employs linear physical programming. The third and fourth models are for scenarios 3 and 4, respectively, and use goal programming and linear physical programming, respectively. The fifth model is for scenario 5 and employs fuzzy goal programming. For the last three models, it is assumed that the manufacturer has incorporated a remanufacturing process into his or her original production system, so that products can be manufactured directly from raw materials or remanufactured from used products (the final demand for the product is met with either new or remanufactured products).

This chapter is organized as follows: Section 8.2 presents the first model. Section 8.3 shows the second model. Section 8.4 presents the third model. Section 8.5 gives the fourth model. Section 8.6 presents the fifth model. Finally, section 8.7 gives some conclusions.

8.2   First Model (Linear Integer Programming)

This model considers a generic reverse supply chain consisting of collection centers, remanufacturing facilities, and demand centers. Optimal transportation of products (used and remanufactured) is achieved using linear integer programming. See [1] for a review of a number of models similar to the one presented in this section. Nomenclature for the model is given in section 8.2.1, the model is formulated in section 8.2.2, and a numerical example is given in section 8.2.3.

8.2.1   Nomenclature

a_1 Space occupied by one unit of remanufactured product

a_2 Space occupied by one unit of used product

CAP_v Capacity of remanufacturing facility v to remanufacture products

C_u Cost per product retrieved at collection center u

d_w Demand of remanufactured products at demand center w

I_{uv} Decision variable representing the number of products to be transported from collection center u to remanufacturing facility v

O_{vw} Decision variable representing the number of products to be transported from remanufacturing facility v to demand center w

R_v Cost of remanufacturing per product at production facility v

S_{1v} Storage capacity of remanufacturing facility v for remanufactured products

S_{2v} Storage capacity of remanufacturing facility v for used products

S_u Storage capacity of collection center u for used products

SUP_u Supply at collection center u

TI_{uv} Cost of transporting one product from collection center u to remanufacturing facility v

TO_{vw} Cost of transporting one product from remanufacturing facility v to demand facility w

u	Collection center
v	Remanufacturing facility
w	Demand center
Y_v	Binary variable (0/1) for selection of recovery facility v

8.2.2 Model Formulation

The following is the single-period and single-product transshipment model formulation that is implemented to achieve minimum overall cost, i.e., sum of the costs of retrieval, inventory, remanufacturing, and transportation of products (used and remanufactured) across the supply chain (in the formulation, it is assumed that the inventory cost of a used product is 25% of its retrieval cost, C_u, and that of a remanufactured product is 25% of its remanufacturing cost, R_v):

Minimize

Retrieval costs
Transportation costs
Remanufacturing costs
Inventory costs

$$\sum_u \sum_v C_u I_{uv} +$$

$$\sum_u \sum_v TI_{uv} I_{uv} + \sum_v \sum_w TO_{vw} O_{vw} +$$

$$\sum_v \sum_w R_v O_{vw} +$$

$$\sum_u \sum_v (C_u / 4).I_{uv} + \sum_v \sum_w (R_v / 4).Q_{vw}$$

(8.1)

subject to

Demand at each demand center must be met

$$\sum_v O_{vw} = d_w ; \forall w$$

(8.2)

Total output of each remanufacturing facility is at most its total input

$$\sum_u I_{uv} \geq \sum_w O_{vw}; \forall v \tag{8.3}$$

Total space occupied by remanufactured products at each remanufacturing facility is at most its capacity for remanufactured products

Total space occupied by used products at each collection center is at most its capacity

$$\sum_w a_1.O_{vw} \leq S_{1v}.Y_v; \forall v \tag{8.4}$$

$$\sum_v a_2.I_{uv} \leq S_u; \forall u \tag{8.5}$$

Total space occupied by used products at each remanufacturing facility is at most its capacity for used products

$$\sum_u a_2.I_{uv} \leq S_{2v}.Y_v; \forall v \tag{8.6}$$

Quantities of transported products are non-negative numbers

$$I_{uv} \geq 0; \forall u, v \tag{8.7}$$

Total output of each remanufacturing facility is at most its capacity to remanufacture

$$O_{vw} \geq 0; \forall v, w \tag{8.8}$$

Total quantity of used products supplied to remanufacturing facilities by each collection center is at most the supply to that collection center

$$\sum_w O_{vw} \leq CAP_v; \forall v \tag{8.9}$$

$$\sum_v I_{uv} \leq SUP_u; \forall u \tag{8.10}$$

Note that a_1 and a_2 are possibly different. The reason is that a component from a used product of one model may be used in a remanufactured product of a different model that may occupy a different amount of space.

Also, the model assumes that there is always enough supply of the used products to satisfy the demand for the remanufactured products and that enough storage space (for used and remanufactured products) is always available at the remanufacturing facilities. Moreover, inventory costs at the collection centers and demand centers are not considered in the model (for example, if the supply of used products is higher than the demand for remanufactured products, the collection centers are deemed to incur inventory costs). However, the model may be revised accordingly in response to revision of existing constraints or addition of new constraints.

8.2.3 Numerical Example

Three collection centers, two remanufacturing facilities, and three demand centers are considered in the example (see figure 8.1). Let the total supply of the used product per period be a triangular fuzzy number (TFN): (600, 650, 700). The defuzzified supply is 633.33 per period (see section 4.4 for the concepts of TFN and defuzzification). Assuming an equal supply rate at all three collection centers, one gets $SUP_1 = SUP_2 = SUP_3 = 211.11$. The other data used for implementation of the model are as follows:

$C_1 = 29$; $C_2 = 25$; $C_3 = 37$; $TI_{1A} = 3$; $TI_{2A} = 4$; $TI_{3A} = 5.3$; $TI_{1B} = 3.2$; $TI_{2B} = 1.4$; $TI_{3B} = 6.7$; $TO_{A1} = 2.6$; $TO_{B1} = 3.2$; $TO_{A2} = 3.4$; $TO_{B2} = 2.5$; $TO_{B3} = 1.6$; $TO_{B3} = 2.1$; $R_A = 4$; $R_B = 4.3$; $d_1 = 100$; $d_2 = 200$; $d_3 = 150$; $a_1 = a_2 = 0.5$; $S_{1A} = 550$; $S_{1B} = 550$; $S_{2A} = 550$; $S_{2B} = 550$; $S_1 = 550$; $S_2 = 550$; $S_3 = 550$; $CAP_A = 300$; $CAP_B = 250$

Upon application of the above data to the model, using LINGO (v4), the following optimal solution is obtained:

$I_{1A} = 211$, i.e., 211 products are to be transported from collection center 1 to recovery facility A

$I_{1B} = 0$, i.e., no products are to be transported from collection center 1 to recovery facility B

$I_{2A} = 0$, i.e., no products are to be transported from collection center 2 to recovery facility A

$I_{2B} = 211$, i.e., 211 products are to be transported from collection center 2 to recovery facility B

FIGURE 8.1
Reverse supply chain (first and second models).

I_{3A} = 28, i.e., 28 products are to be transported from collection center 3 to recovery facility A

I_{3B} = 0, i.e., no products are to be transported from collection center 3 to recovery facility B

O_{A1} = 100, i.e., 100 products are to be transported from recovery facility A to demand center 1

O_{A2} = 0, i.e., no products are to be transported from recovery facility A to demand center 2

O_{A3} = 139, i.e., 139 products are to be transported from recovery facility A to demand center 3

O_{B1} = 0, i.e., no products are to be transported from recovery facility B to demand center 1

O_{B2} = 200, i.e., 200 products are to be transported from recovery facility B to demand center 2

O_{B3} = 11, i.e., 11 products are to be transported from recovery facility B to demand center 3

8.3   Second Model (Linear Physical Programming)

This model considers a generic reverse supply chain consisting of collection centers, remanufacturing facilities, and demand centers. Optimal transportation of products (used and remanufactured) is achieved using linear physical programming. Section 8.3.1 presents the model formulation (using the nomenclature given in section 8.2.1), and section 8.3.2 gives a numerical example.

8.3.1   Model Formulation

Class 1H criteria

Total retrieval cost per period (h_1) given by

$$h_1 = \sum_u \sum_v C_u I_{uv}$$

(8.11)

Class 1S criteria (smaller is better)

Total transportation cost per period (g_1) given by

$$g_1 = \sum_u \sum_v TI_{uv} I_{uv} + \sum_v \sum_w TO_{vw} O_{vw}$$

(8.12)

Total remanufacturing cost per period (g_2) given by

$$g_2 = \sum_v \sum_w R_v O_{vw} \tag{8.13}$$

Total inventory cost per period (g_3) given by

$$g_3 = \sum_u \sum_v (C_u / 4).I_{uv} + \sum_v \sum_w (R_v / 4).Q_{vw} \tag{8.14}$$

Goal constraints

$h_1 \leq \text{RETMAX}$ (retrieval cost is not more than maximum allowed value—RETMAX) (8.15)

$g_p - d_{pr}^+ \leq t_{p(r-1)}^+$ (deviation is measured from corresponding target value) (8.16)

$g_p \leq t_{p5}^+$ (criterion value is in acceptable range) (8.17)

$d_{pr}^+ \geq 0;$ (deviation is a nonnegative number) (8.18)

System constraints

$\sum_v O_{vw} = d_w \forall w$ (demand at each demand center must be met) (8.19)

$\sum_u I_{uv} = SUP_u \forall v$ (all products must be transported from each collection center) (8.20)

$\sum_w O_{vw} = \sum_u I_{uv} \forall v$ (number of remanufactured products is equal to number of used ones) (8.21)

$$\sum_u a_2.I_{uv} \le S_{2v}; \forall v \text{ (space occupied by used products is at most capacity) (8.22)}$$

$$I_{uv} \ge 0 \forall u,v \quad \text{(quantities of used products are nonnegative numbers) (8.23)}$$

$$O_{vw} \ge 0 \forall v,w \quad \text{(quantities of remanufactured products are nonnegative numbers)} \tag{8.24}$$

It should be noted here that more constraints may be added to the above model, as desired by the decision maker.

8.3.2   Numerical Example

Consider the reverse supply chain shown in figure 8.1 in this example as well. The data used for implementation of the model are as follows:

$C_1 = 0.1; C_2 = 0.1; C_3 = 0.2; TI_{1A} = 0.02; TI_{1B} = 0.1; TI_{2A} = 0.2; TI_{2B} = 3; TI_{3A} = 0.1;$
$TI_{3B} = 1.2; TO_{A1} = 4; TO_{A2} = 5; TO_{A3} = 0.04; TO_{B1} = 1; TO_{B2} = 0.1; TO_{B3} = 0.2; R_A$
$= 0.2; R_B = 0.3; d_1 = 90; d_2 = 80; d_3 = 80; SUP_1 = 75; SUP_2 = 150; SUP_3 = 25; a_1 =$
$a_2 = 0.5; S_{2A} = S_{2C} = 400$

Furthermore, the target values for each soft criterion are shown in table 8.1, and the incremental weights obtained by the LPP weight algorithm [2] are shown in table 8.2.

Upon application of the above data to the model, using LINGO (v4), the following optimal solution is obtained:

TABLE 8.1
Preference table (second model)

Criteria	$t_{p1}+$	$t_{p2}+$	$t_{p3}+$	$t_{p4}+$	$t_{p5}+$
g_1	100	200	300	400	500
g_2	150	250	290	450	600
g_3	70	150	250	300	450

TABLE 8.2
Output of LPP weight algorithm (second model)

Criteria	$\Delta w_{p2}+$	$\Delta w_{p3}+$	$\Delta w_{p4}+$	$\Delta w_{p5}+$
g_1	0.025	0.085	0.132	0.024
g_2	0.017	0.011	0.026	0.479
g_3	0.013	0.031	0.881	0.012

$I_{1A} = 0; I_{2A} = 80; I_{3A} = 0; I_{1B} = 75; I_{2B} = 70; I_{3B} = 25; O_{A1} = 0; O_{A2} = 0; O_{A3} = 80;$
$O_{B1} = 90; O_{B2} = 80; O_{B3} = 0$

Interpretation of this solution could be made in a way similar to that in section 8.2.3.

8.4 Third Model (Goal Programming)

This model considers a generic closed-loop supply chain consisting of collection centers, production facilities, and demand centers. Goal programming is employed to address the following issues in one continuous phase, besides optimal transportation of products: selection of used products and evaluation of production facilities. Section 8.4.1 gives the nomenclature for the model, section 8.4.2 presents the formulation of the model, and section 8.4.3 illustrates the model with a numerical example.

8.4.1 Nomenclature

A_{iuv}	Decision variable representing number of used products of type i transported from collection center u to production facility v
B_{ivw}	Decision variable representing number of new/remanufactured products of type i transported from production facility v to demand center w
b_i	Probability of breakage of product i
TA_{uv}	Cost to transport one product from collection center u to production facility v
TB_{vw}	Cost to transport one product from production facility v to demand center w
CC_u	Cost per product retrieved at collection center u
CNP_v	Cost to produce one unit of new product at production facility v
CR_v	Cost to produce one unit of remanufactured product at production facility v
C_{di}	Disposal cost of product i
DI_i	Disposal cost index of product i (0 = lowest, 10 = highest)
DT_i	Disassembly time for product i
DC	Disassembly cost/unit time
i	Product type

MINTPS	Minimum throughput per supply
N_{ivw}	Decision variable representing number of new products of type i transported from production facility v to demand center w
Nd_{iw}	Net demand for product type i (remanufactured or new) at demand center w
PRC_i	Percent of recyclable contents by weight in product i
$RCYR_i$	Total recycling revenue of product i
RSR_i	Total resale revenue of product i
$RCRI_i$	Recycling revenue index of product i
S_{1v}	Storage capacity of production facility v for used products
S_{2v}	Storage capacity of production facility v for remanufactured and new products
S_u	Storage capacity of collection center u
SP_i	Selling price of one unit of new product of type i
SU_{iu}	Supply of used product i at collection center u
SF_v	Supply of used products that are fit for remanufacturing (viz., excluding products selected for recycling or disposal) and new products, at production facility v
TP_v	Throughput (considering only remanufactured products) of production facility v
u	Collection center
v	Production facility
w	Demand center
W_i	Weight of product i
x_1	Space occupied by one unit of used product
x_2	Space occupied by one unit of remanufactured or new product
Y_v	Decision variable signifying selection of production facility v (1 if selected, 0 if not)
Z_{iu}	Decision variable representing number of units of product type i picked for remanufacturing at collection center u ($SU_{iu} - Z_{iu}$ = recycled or disposed)
δ_v	Factor that accounts for unassignable causes of variations at production facility v

8.4.2 Model Formulation

This goal programming model, in one continuous phase, determines the number of used products of each type to be picked for remanufacturing, identifies efficient production facilities, and achieves transportation of the right quantities of products (used, remanufactured, and new) across a closed-loop supply chain.

It is assumed that the inventory costs of a used product and a remanufactured product are 20% of the collection and remanufacturing costs, respectively, and that of a newly produced product is 25% of the production cost.

The following three goals are considered:

1. Maximize the total profit in the supply chain (TP).
2. Maximize the revenue from recycling (RR).
3. Minimize the number of disposed items (NDIS).

The first two goals involve minimizing the negative deviation from the respective target values, whereas the third goal, which has an *environmentally benign* focus rather than a *financial* focus, involves minimizing the positive deviation from the target value.

The revenue and cost criteria and the system constraints considered in the model are:

Revenues

1. Reuse revenue:

$$\sum_i \sum_u \{Z_{iu} \times RSR_i\} \tag{8.25}$$

2. Recycle revenue:

$$\sum_i \sum_u \{(SU_{iu} - Z_{iu}) \times RCRI_i \times W_i \times PRC_i\} \tag{8.26}$$

3. New product sale revenue:

$$\sum_i \sum_v \sum_w SP_i \times N_{ivw} \tag{8.27}$$

Costs

1. Collection/retrieval cost:

$$\sum_u \sum_i CC_u \times SU_{iu} \tag{8.28}$$

2. Processing cost = disassembly cost of used products + remanufacturing cost of used products + new products' production cost:

$$\left(DC \times \sum_i \sum_u \sum_v DT_i \times A_{iuv} \right) + \sum_i \sum_u \sum_v CR_v \times B_{ivw} + \sum_i \sum_v \sum_w CNP_v \times N_{ivw}$$

(8.29)

3. Inventory cost = cost of carrying used products' inventory at collection centers + cost of carrying remanufactured products' inventory at production facilities + cost of carrying newly manufactured products' inventory at production facilities:

$$\sum_i \sum_u \sum_v (CC_u / 5) \times A_{iuv} + \left(\sum_i \sum_v \sum_w \{ (CR_v / 4) \times B_{ivw} + (CNP_v / 4) * N_{ivw} \} \right)$$

(8.30)

4. Transportation cost = cost of transporting used products from collection centers to production facilities + cost of transporting remanufactured and new products from production facilities to demand centers:

$$TA_{uv} \times \sum_i \sum_u \sum_v A_{iuv} + TB_{vw} \times \sum_i \sum_v \sum_w (B_{ivw} + N_{ivw})$$

(8.31)

5. Disposal cost: Apart from the number of units disposed of and the cost to dispose of a product, disposal cost depends on the percentage of recyclable content in the product, and the disposal cost index (a number on a scale 0–10; the higher the number, the more difficult or expensive it is to dispose of the product):

$$\sum_i \sum_u \{ (SU_{iu} - Z_{iu}) \times DI_i \times W_i \times (1 - PRC_i) \} \times C_{di}$$

(8.32)

System constraints

1. The number of used products sent to all production facilities from collection center u must be equal to the number of used products picked for remanufacturing at that collection center:

$$\sum_v A_{iuv} = Z_{iu}$$

(8.33)

2. The demand at each center w must be met by either new or remanufactured products:

$$\sum_v (B_{ivw} + N_{ivw}) = Nd_{iw} \forall w \qquad (8.34)$$

3. The number of remanufactured products transported from production facility v to demand center w equals (number of used products fit for remanufacturing that are transported from collection center u to that production facility) multiplied by δ_v. δ_v is a factor that accounts for the unassignable (common) causes of variation at production facility v. That is, there is no loss of products in the supply chain due to reasons other than common cause variations over which there is no control:

$$\sum_w B_{ivw} = \sum_u A_{iuv} \times \delta_v \forall v \qquad (8.35)$$

4. The number of used products of type i picked for remanufacturing at collection center u must be at most equal to the total number of used products fit for remanufacturing:

$$Z_{iu} \leq SU_{iu}(1 - b_i) \qquad (8.36)$$

5. The total number of used products of all types collected (before accounting for probability of breakage) at all collection centers must be at least equal to the net demand (this is to encourage the use of remanufactured products):

$$\sum_i \sum_u SU_{iu} \geq \sum_i \sum_w Nd_{iw} \qquad (8.37)$$

6. The number of remanufactured products must be at most equal to the net demand (this is to avoid excess remanufacturing):

$$\sum_i \sum_u Z_{iu} \leq \sum_i \sum_w Nd_{iw} \qquad (8.38)$$

7. The space occupied by used products at production facility v must at most be equal to the space available for used products at that facility:

$$x_1 \sum_i \sum_u A_{iuv} \leq S_{1v} . Y_v$$

$$(8.39)$$

8. The space occupied by new and remanufactured products at production facility v must at most be equal to the space available for new and remanufactured products at that production facility (assuming both new and remanufactured products occupy the same space):

$$\sum_i \sum_w x_2 (B_{ivw} + N_{ivw}) \leq S_{2v} \times Y_v$$

$$(8.40)$$

9. The space occupied by used products at collection center u must at most be equal to the space available for used products at that collection center:

$$x_1 \sum_i \sum_v a_{iuv} \leq S_u \qquad (8.41)$$

10. The ratio of throughput to supply of used products of a production facility must at least be equal to a preset value for the production facility to be considered efficient (this is valid only for remanufactured products):

$$\left(TP_v / SF_v \right) Y_v \geq MINTPS \qquad (8.42)$$

11. Nonnegativity constraints:

$$A_{iuv}, B_{ivw}, N_{ivw}, Z_{iu} \geq 0, \forall u, v, w, i \qquad (8.43)$$

12. $Y_v \in [0,1] \forall v$, 0 if facility v not selected, 1 if selected $\qquad (8.44)$

8.4.3 Numerical Example

Three collection centers, two production facilities, two demand centers, and three types of products are considered in the example (see figure 8.2). The data used to implement the goal programming model are:

$CC_u = 0.01$; $SU_{11} = 50$; $SU_{12} = 45$; $SU_1 = 25$; $SU_{21} = 35$; $SU_{22} = 38$; $SU_{23} = 22$; $SU_{31} = 30$; $SU_{32} = 35$; $SU_{33} = 28$; $DC = 0.05$; $DT_1 = 10$; $DT_2 = 12$; $DT_3 = 9$; $CR_1 = 13$; $CR_2 = 10$; $CNP_1 - 60$; $CNP_2 = 45$; $TA_{11} = 0.01$; $TA_{12} = 0.09$; $TA_{21} = 0.5$; $TA_{22} = 0.1$; TA_{31}

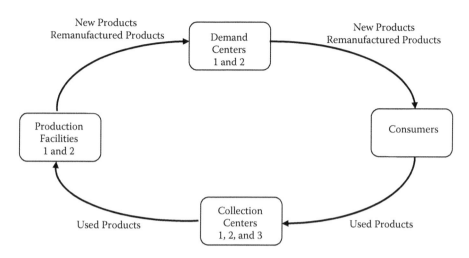

FIGURE 8.2
Closed-loop supply chain (third, fourth, and fifth models).

$= 0.02$; $TA_{32} = 0.04$; $TB_{11} = 0.04$; $TB_{12} = 0.03$; $TB_{21} = 0.09$; $TB_{22} = 0.05$; $DI_1 = 4$; $DI_2 = 6$; $DI_3 = 5$; $W_1 = 0.8$; $W_2 = 1.0$; $W_3 = 0.9$; $PRC_1 = 0.5$; $PRC_2 = 0.6$; $PRC_3 = 0.75$; $Cd_1 = 0.2$; $Cd_2 = 0.5$; $Cd_3 = 0.3$; $RSR_1 = 30$; $RSR_2 = 40$; $RSR_3 = 45$; $RCYR_1 = 1.5$; $RCYR_2 = 2$; $RCYR_3 = 2.5$; $RCRI_1 = 7$; $RCRI_2 = 4$; $RCRI_3 = 5$; $SP_1 = 65$; $SP_2 = 55$; $SP_3 = 60$; $Nd_{11} = 20$; $Nd_{12} = 15$; $Nd_{21} = 16$; $Nd_{22} = 22$; $Nd_{31} = 25$; $Nd_{32} = 20$; $\delta_1 = 0.4$; $\delta_2 = 0.6$; $b_1 = 0.2$; $b_2 = 0.4$; $b_3 = 0.3$; $X_1 = 0.7$; $S_{11} = 400$; $S_{12} = 400$; $S_1 = 150$; $S_2 = 150$; $S_3 = 150$; $X_2 = 0.7$; $S_{21} = 500$; $S_{22} = 500$; $MINTPS = 0.25$

Upon application of the above data to the model, using LINGO (v4), the following optimal solution is obtained:

$TP = 3{,}945$ (target $= 2{,}500$); $RR = 951$ (target $= 750$); $NDIS = 74$ (target $= 50$); $Z_{12} = 35$; $Z_{21} = 2$; $Z_{22} = 23$; $Z_{23} = 13$; $Z_{31} = 20$; $Z_{32} = 12$; $Z_{33} = 12$; $N_{111} = 3$; $N_{112} = 5$; $N_{211} = 8$; $N_{222} = 1$; $N_{211} = 3$; $N_{312} = 14$; $A_{121} = 15$; $A_{122} = 20$; $A_{212} = 2$; $A_{221} = 5$; $A_{222} = 18$; $A_{231} = 5$; $A_{232} = 8$; $A_{311} = 11$; $A_{312} = 9$; $A_{321} = 12$; $A_{331} = 3$; $B_{111} = 8$; $B_{112} = 4$; $B_{211} = 5$; $B_{212} = 3$; $B_{311} = 15$; $B_{312} = 6$; $B_{121} = 9$; $B_{122} = 6$; $B_{221} = 3$; $B_{222} = 18$; $B_{321} = 7$; $Y_1 = 1$; $Y_2 = 1$

It is evident from the above solution that both of the production facilities are chosen for the network design. Also, 71% of the net demand is satisfied by remanufactured products and the remaining 29% by newly manufactured products.

8.5   Fourth Model (Linear Physical Programming)

This model considers a generic closed-loop supply chain consisting of collection centers, production facilities, and demand centers. Linear physical programming is employed to address the following issues in one continuous phase, besides optimal transportation of products: selection of used products and evaluation of production facilities. Section 8.5.1 presents the model formulation using the nomenclature given in section 8.4.1, and section 8.5.2 gives a numerical example.

8.5.1   Model Formulation

This linear physical programming model, in one continuous phase, determines the number of used products of each type to be picked for remanufacturing, identifies efficient production facilities, and achieves transportation of the right quantities of products (used, remanufactured, and new) across a closed-loop supply chain.

It is assumed that the inventory costs of a used product and a remanufactured product are 20% of the collection and remanufacturing costs, respectively and that the inventory cost of a newly produced product is 25% of the production cost.

The revenue and cost criteria and system constraints considered in the model are:

Costs: Class 1S (smaller is better)

  1. Collection/retrieval cost:

$$\sum_{u}\sum_{i} CC_u \times SU_{iu}$$

(8.45)

  2. Processing cost = disassembly cost of used products + remanufacturing cost of used products + new products' production cost:

$$\left(DC \times \sum_{i}\sum_{u}\sum_{v} DT_i \times A_{iuv} \right) + \sum_{i}\sum_{u}\sum_{v} CR_v \times B_{ivw} + \sum_{i}\sum_{v}\sum_{w} CNP_v \times N_{ivw}$$

(8.46)

  3. Transportation cost = cost of transporting used products from collection centers to production facilities + cost of transporting remanufactured and new products from production facilities to demand centers:

$$TA_{uv} \times \sum_i \sum_u \sum_v A_{iuv} + TB_{vw} \times \sum_i \sum_v \sum_w (B_{ivw} + N_{ivw})$$

(8.47)

4. Disposal cost: Apart from the number of units disposed of and the cost to dispose of a product, disposal cost depends on the percentage of recyclable content in the product and the disposal cost index (a number on a scale 0–10; the higher the number, the more difficult or expensive it is to dispose of the product):

$$\sum_i \sum_u \{(SU_{iu} - Z_{iu}) \times DI_i \times W_i \times (1 - PRC_i)\} \times C_{di}$$

(8.48)

Revenues: Class 2S (larger is better)

1. Reuse revenue:

$$\sum_i \sum_u \{Z_{iu} \times RSR_i\}$$

(8.49)

2. Recycle revenue:

$$\sum_i \sum_u \{(SU_{iu} - Z_{iu}) \times RCRI_i \times W_i \times PRC_i\}$$

(8.50)

3. New product sales revenue:

$$\sum_i \sum_v \sum_w SP_i \times N_{ivw}$$

(8.51)

System constraints

1. The number of used products sent to all production facilities from collection center u must be equal to the number of used products picked for remanufacturing at that collection center:

$$\sum_v A_{iuv} = Z_{iu}$$

(8.52)

2. The demand at each center w must be met by either new or remanufactured products:

$$\sum_v (B_{ivw} + N_{ivw}) = Nd_{iw} \forall w$$

(8.53)

3. The number of remanufactured products transported from production facility v to demand center w equals (number of used products fit for remanufacturing that are transported from collection center u to that production facility) multiplied by δ_v. δ_v is a factor that accounts for the unassignable (common) causes of variation at production facility v. That is, there is no loss of products in the supply chain due to reasons other than common cause variations, over which there is no control:

$$\sum_w B_{ivw} = \sum_u A_{iuv} \times \delta_v \forall v \tag{8.54}$$

4. The number of used products of type i picked for remanufacturing at collection center u must be at most equal to the total number of used products fit for remanufacturing:

$$Z_{iu} \leq SU_{iu}(1 - b_i) \tag{8.55}$$

5. The total number of used products of all types collected (before accounting for probability of breakage) at all collection centers must be at least equal to the net demand (this is to encourage the use of remanufactured products):

$$\sum_i \sum_u SU_{iu} \geq \sum_i \sum_w Nd_{iw} \tag{8.56}$$

6. The number of remanufactured products must be at most equal to the net demand (this is to avoid excess remanufacturing):

$$\sum_i \sum_u Z_{iu} \leq \sum_i \sum_w Nd_{iw} \tag{8.57}$$

7. The space occupied by used products at production facility v must at most be equal to the space available for used products at that facility:

$$x_1 \sum_i \sum_u A_{iu} \leq S_{1v}.Y_v \tag{8.58}$$

8. The space occupied by new and remanufactured products at production facility v must at most be equal to the space available

for new and remanufactured products at that production facility (assuming both new and remanufactured products occupy the same space):

$$\sum_i \sum_w x_2 (B_{ivw} + N_{ivw}) \le S_{2v} \times Y_v \tag{8.59}$$

9. The space occupied by used products at collection center u must at most be equal to the space available for used products at that collection center:

$$x_1 \sum_i \sum_v a_{iuv} \le S_u \tag{8.60}$$

10. The ratio of throughput to supply of used products of a production facility must at least be equal to a preset value for the production facility to be considered efficient (this is valid only for remanufactured products):

$$\left(TP_v / SF_v \right) Y_v \ge MINTPS \tag{8.61}$$

11. Nonnegativity constraints:

$$A_{iuv}, B_{ivw}, N_{ivw}, Z_{iu} \ge 0, \forall\, u, v, w, i \tag{8.62}$$

12. $Y_v \in [0,1]\, \forall\, v$, 0 if facility v not selected, 1 if selected $\tag{8.63}$

8.5.2 Numerical Example

Three collection centers, two production facilities, two demand centers, and three types of products are considered in the example (see figure 8.2). The data used to implement the linear physical programming model are:

$CC_u = 0.01$; $SU_{11} = 20$; $SU_{12} = 25$; $SU_{13} = 15$; $SU_{21} = 25$; $SU_{22} = 18$; $SU_{23} = 15$; $SU_{31} = 17$; $SU_{32} = 9$; $SU_{33} = 15$; $DC = 0.005$; $DT_1 = 10$; $DT_2 = 12$; $DT_3 = 9$; $CR_1 = 10$; $CR_2 = 8$; $CNP_1 = 55$; $CNP_2 = 65$; $TA_{11} = 0.001$; $TA_{12} = 0.009$; $TA_{21} = 0.01$; $TA_{22} = 0.002$; $TA_{31} = 0.004$; $TA_{32} = 0.003$; $TB_{11} = 0.004$; $TB_{12} = 0.003$; $TB_{21} = 0.009$; $TB_{22} = 0.005$; $DI_1 = 4$; $DI_2 = 6$; $DI_3 = 5$; $W_1 = 0.8$; $W_2 = 1.0$; $W_3 = 0.9$; $PRC_1 = 0.65$; $PRC_2 = 0.6$; $PRC_3 = 0.75$; $Cd_1 = 0.02$; $Cd_2 = 0.05$; $Cd_3 = 0.03$; $RSR_1 = 80$; $RSR_2 = 80$; $RSR_3 = 65$; $RCYR_1 = 5$; $RCYR_2 = 7$; $RCYR_3 = 105$; $RCRI_1 = 7$; $RCRI_2 = 4$; $RCRI_3 = 6$; $SP_1 = 100$; $SP_2 = 110$; $SP_3 = 95$; $Nd_{11} = 20$; $Nd_{12} = 15$; $Nd_{21} = 16$; $Nd_{22} = 22$; $Nd_{31} = 25$;

TABLE 8.3

Target values of criteria (fourth model)

Criteria	$t_{p1}+$	$t_{p2}+$	$t_{p3}+$	$t_{p4}+$	$t_{p5}+$
g_1	1	3	5	7	9
g_2	5	10	13	17	20
g_3	2	2.5	5	7	10
g_4	2	5	7	9	13
Criteria	$t_{p1}-$	$t_{p2}-$	$t_{p3}-$	$t_{p4}-$	$t_{p5}-$
g_5	10	15	20	25	30
g_6	10	15	17	19	22
g_7	15	17	20	25	35

TABLE 8.4

Output of LPP weight algorithm (fourth model)

Criteria	$\Delta w_{p2}+$	$\Delta w_{p3}+$	$\Delta w_{p4}+$	$\Delta w_{p5}+$	$\Delta w_{p2}-$	$\Delta w_{p3}-$	$\Delta w_{p4}-$	$\Delta w_{p5}-$
g_1	0.05	0.17	0.75	3.29	—	—	—	—
g_2	0.02	0.13	0.34	2.34	—	—	—	—
g_3	0.20	0.02	1.34	4.29	—	—	—	—
g_4	0.03	0.25	1.29	2.82	—	—	—	—
g_5	—	—	—	—	0.02	0.02	0.05	0.12
g_6	—	—	—	—	0.02	0.09	0.13	0.11
g_7	—	—	—	—	0.05	0.03	0.04	0.02

$Nd_{32} = 20$; $\delta_1 = 0.85$; $\delta_2 = 0.75$; $b_1 = 0.2$; $b_2 = 0.4$; $b_3 = 0.3$; $X_1 = 0.7$; $S_{11} = 400$; $S_{12} = 400$; $S_1 = 150$; $S_2 = 150$; $S_3 = 150$; $X_2 = 0.7$; $S_{21} = 500$; $S_{22} = 500$; $MINTPS = 0.25$

The target values for the criteria are given in table 8.3 (target values are scaled by a factor of 10), and table 8.4 shows the incremental weights obtained using the LPP weight algorithm [2].

Upon application of the above data to the model, using LINGO (v4), the following optimal solution is obtained:

$Z_{11} = 0$; $Z_{12} = 0$; $Z_{13} = 0$; $Z_{21} = 15$; $Z_{22} = 11$; $Z_{23} = 9$; $Z_{31} = 3$; $Z_{32} = 6$; $Z_{33} = 6$; $N_{111} = 15$; $N_{112} = 11$; $N_{211} = 0$; $N_{212} = 0$; $N_{311} = 18$; $N_{312} = 15$; $N_{121} = 5$; $N_{122} = 4$; $N_{221} = 4$; $N_{222} = 5$; $N_{321} = 0$; $N_{322} = 5$; $A_{111} = 0$; $A_{112} = 0$; $A_{121} = 0$; $A_{122} = 0$; $A_{131} = 0$; $A_{132} = 0$; $A_{211} = 11$; $A_{212} = 4$; $A_{221} = 8$; $A_{222} = 3$; $A_{231} = 7$; $A_{232} = 2$; $A_{311} = 3$; $A_{312} = 0$; $A_{321} = 5$; $A_{322} = 2$; $A_{331} = 0$; $A_{332} = 0$; $B_{111} = 0$; $B_{112} = 0$; $B_{211} = 8$; $B_{212} = 14$; $B_{311} = 6$; $B_{312} = 0$; $B_{121} = 0$; $B_{122} = 0$; $B_{221} = 4$; $B_{222} = 3$; $B_{321} = 1$; $B_{322} = 0$; $Y_1 = 1$; $Y_2 = 1$

It is evident from the above solution that both production facilities are chosen for the network design. Also, 69% of the net demand is satisfied by newly manufactured products and the remaining 31% by remanufactured products. In addition, demand for the first product type is completely satisfied by newly manufactured products of that type.

8.6   Fifth Model (Fuzzy Goal Programming)

This model considers a generic closed-loop supply chain consisting of collection centers, production facilities, and demand centers. Fuzzy goal programming is employed to address the following issues in one continuous phase, besides optimal transportation of products: selection of used products and evaluation of production facilities. Section 8.6.1 presents the model formulation using the nomenclature given in section 8.4.1, and section 8.6.2 gives a numerical example.

8.6.1   Model Formulation

This fuzzy goal programming model, in one continuous phase, determines the number of used products of each type to be picked for remanufacturing, identifies efficient production facilities, and achieves transportation of the right quantities of products (used, remanufactured, and new) across a closed-loop supply chain.

It is assumed that the inventory costs of a used product and a remanufactured product are 20% of the collection and remanufacturing costs, respectively, and that the inventory cost of a newly produced product is 25% of the production cost.

Three goals are considered in the model:

1. Maximize the total profit in the closed-loop supply chain (TP).
2. Maximize the revenue from recycling (RR).
3. Minimize the number of disposed items (NDIS).

The first two goals involve minimizing the negative deviation from the respective target values, whereas the third goal, which has an *environmentally benign* character rather than a *financial* basis, involves minimizing the positive deviation from the target value. The membership functions for the three goals are

$$\mu_1 = \begin{cases} 1 & \text{if } TP \geq TP^* \\[2mm] \dfrac{TP - TP_L}{TP^* - TP_L} & \text{if } TP_L \leq TP \leq TP^* \\[2mm] 0 & \text{if } TP \leq TP_L \end{cases}$$

$$(8.64)$$

where TP^* is the aspiration level of total profit (TP) and TP_L is the lower tolerance level of TP;

$$
\mu_2 = \begin{cases} 1 & \text{if } RR \geq RR^* \\ \dfrac{RR - RR_L}{RR^* - RR_L} & \text{if } RR_L \leq RR \leq RR^* \\ 0 & \text{if } RR \leq RR_L \end{cases}
\tag{8.65}
$$

where RR^* is the aspiration level of recycling revenue (RR) and RR_L is the lower tolerance level of RR; and

$$
\mu_3 = \begin{cases} 1 & \text{if } NDI \leq NDI^* \\ \dfrac{NDI_U - NDI}{NDI_U - NDI^*} & \text{if } NDI^* \leq NDI \leq NDI_U \\ 0 & \text{if } NDI \geq NDI_U \end{cases}
\tag{8.66}
$$

where NDI^* is the aspiration level of number of disposed items (NDI) and NDI_U is the upper tolerance level of NDI.

According to the concept of Fibonacci numbers, starting with 1 and 2, the weight values for μ_1, μ_2, and μ_3 are 0.5, 0.33, and 0.17. The objective function for the weighted fuzzy goal programming model is as follows:

$$
Maximize\ V(\mu) = (0.5 \times \mu_1) + (0.33 \times \mu_2) + (0.17 \times \mu_3)
\tag{8.67}
$$

The cost and revenue criteria and the system constraints considered in the model include:

Revenues

1. Reuse revenue:

$$
\sum_i \sum_u \{Z_{iu} \times RSR_i\}
\tag{8.68}
$$

2. Recycle revenue:

$$
\sum_i \sum_u \{(SU_{iu} - Z_{iu}) \times RCYI_i \times W_i \times PRC_i\}
\tag{8.69}
$$

3. New product sales revenue:

$$\sum_i \sum_v \sum_w SP_i \times N_{ivw} \qquad (8.70)$$

Costs

1. Collection/retrieval cost:

$$\sum_u \sum_i CC_u \times SU_{iu} \qquad (8.71)$$

2. Processing cost = disassembly cost of used products + remanufacturing cost of used products + new products production cost in the forward supply chain:

$$\left(DC \times \sum_i \sum_u \sum_v DT_i \times A_{iuv} \right) + \sum_i \sum_u \sum_v CR_v \times B_{ivw} + \sum_i \sum_v \sum_w CNP_v \times N_{ivw} \qquad (8.72)$$

3. Inventory cost = cost of carrying used products inventory at the collection center + cost of carrying remanufactured products inventory at the production facility + cost of carrying newly manufactured products inventory at the production facility:

$$\sum_i \sum_u \sum_v (CC_u / 5) \times A_{iuv} + \left(\sum_i \sum_v \sum_w \{(CR_v / 4) \times B_{ivw} + (CNP_v / 4) \times N_{ivw}\} \right) \qquad (8.73)$$

4. Transportation costs = cost of transporting used products from collection centers to production facility + remanufactured and new products from production facilities to demand centers:

$$TA_{uv} \times \sum_i \sum_u \sum_v A_{iuv} + TB_{vw} \times \sum_i \sum_v \sum_w (B_{ivw} + N_{ivw}) \qquad (8.74)$$

5. Disposal cost: Apart from the number of units disposed of and the cost to dispose of a product, disposal cost depends on the percentage of recyclable content in the product and the disposal cost index (a number on a scale 0–10; the higher the number, the more difficult or expensive it is to dispose of the product):

$$\sum_{i}\sum_{u}\{(SU_{iu} - Z_{iu}) \times DI_i \times W_i \times (1 - PRC_i)\} \times C_{di} \qquad (8.75)$$

System constraints

1. The number of used products sent to all production facilities from a collection center u must be equal to the number of used products picked for remanufacturing at that collection center:

$$\sum_{v} A_{iuv} = Z_{iu} \qquad (8.76)$$

2. The demand at each center w must be met by either new or remanufactured products:

$$\sum_{v} (B_{ivw} + N_{ivw}) = Nd_{iw} \ \forall \ w \qquad (8.77)$$

3. The number of remanufactured products transported from a production facility v to a demand center w equals (number of used products fit for remanufacturing that are transported from collection center u to that production facility) multiplied by δ_v v, i.e., no loss of products in the supply chain due to reasons other than common cause variations, over which there is no control. δ_v is a factor that accounts for the unassignable (common) causes of variation at the production facility v:

$$\sum_{w} B_{ivw} = \sum_{u} A_{iuv} \times \delta_v \ \forall \ v \qquad (8.78)$$

4. The total number of used products of type i picked for remanufacturing at collection center u must be at most equal to the total number of used products fit for remanufacturing:

$$Z_{iu} \leq SU_{iu}(1 - b_i) \qquad (8.79)$$

5. The total number of used products of all types collected (before accounting for probability of breakage) at all collection centers must be at least equal to the net demand (this is to encourage the use of remanufactured products):

$$\sum_{i}\sum_{u} SU_{iu} \geq \sum_{i}\sum_{w} Nd_{i_1} \tag{8.80}$$

6. The number of remanufactured products must be at most equal to the net demand (this is to avoid excess remanufacturing):

$$\sum_{i}\sum_{u} Z_{iu} \leq \sum_{i}\sum_{w} Nd_{i} \tag{8.81}$$

7. The space occupied by used products at the production facility, v, must at most be equal to the space available for used products at that facility:

$$x_1 \sum_{i}\sum_{u} A_{iuv} \leq S_{1v}.Y_v \tag{8.82}$$

8. The space occupied by new and remanufactured products at the production facility, v, must at most be equal to the space available for new and remanufactured products at that production facility (assuming both new and remanufactured products occupy the same space):

$$\sum_{i}\sum_{w} x_2 (B_{ivw} + N_{ivw}) \leq S_{2v} \times Y_v \tag{8.83}$$

9. The space occupied by used products at the collection center, u, must at most be equal to the space available for used products at that collection center:

$$x_1 \sum_{i}\sum_{v} a_{iuv} \leq S_u \tag{8.84}$$

10. The ratio of throughput to supply of used products of a production facility must at least be equal to a preset value for the production facility to be considered efficient (this is valid only for remanufactured products):

$$\left(TP_v / SF_v \right) Y_v \geq MINTPS \tag{8.85}$$

11. Nonnegativity constraints:

$$A_{iuv}, B_{ivw}, N_{ivw}, Z_{iu} \geq 0, \forall\, u, v, w, i \tag{8.86}$$

12. $Y_v \in [0,1]\,\forall\, v$, 0 if facility v not selected, 1 if selected   (8.87)

8.6.2   Numerical Example

Three collection centers, two production facilities, two demand centers, and three types of products are considered in the example (see figure 8.2). The data used to implement the fuzzy goal programming model are:

$CC_u = 0.01$; $SU_{11} = 50$; $SU_{12} = 45$; $SU_1 = 25$; $SU_{21} = 35$; $SU_{22} = 38$; $SU_{23} = 22$; $SU_{31} = 30$; $SU_2 = 35$; $SU_{33} = 28$; $DC = 0.05$; $DT_1 = 10$; $DT_2 = 12$; $DT_3 = 9$; $CR_1 = 13$; $CR_2 = 10$; $CNP_1 = 60$; $CNP_2 = 45$; $TA_{11} = 0.01$; $TA_{12} = 0.09$; $TA_{21} = 0.5$; $TA_{22} = 0.1$; $TA_{31} = 0.02$; $TA_{32} = 0.04$; $TB_{11} = 0.04$; $TB_{12} = 0.03$; $TB_{21} = 0.09$; $TB_{22} = 0.05$; $DI_1 = 4$; $DI_2 = 6$; $DI_3 = 5$; $W_1 = 0.8$; $W_2 = 1.0$; $W_3 = 0.9$; $PRC_1 = 0.5$; $PRC_2 = 0.6$; $PRC_3 = 0.75$; $Cd_1 = 0.2$; $Cd_2 = 0.5$; $Cd_3 = 0.3$; $RSR_1 = 30$; $RSR_2 = 40$; $RSR_3 = 45$; $RCYR_1 = 1.5$; $RCYR_2 = 2$; $RCYR_3 = 2.5$; $RCRI_1 = 7$; $RCRI_2 = 4$; $RCRI_3 = 5$; $SP_1 = 65$; $SP_2 = 53$; $SP_3 = 60$; $Nd_{11} = 20$; $Nd_{12} = 15$; $Nd_{21} = 16$; $Nd_{22} = 22$; $Nd_{31} = 25$; $Nd_{32} = 20$; $\delta_1 = 0.4$; $\delta_2 = 0.6$; $b_1 = 0.2$; $b_2 = 0.4$; $b_3 = 0.3$; $X_1 = 0.7$; $S_{11} = 400$; $S_{12} = 400$; $S_1 = 150$; $S_2 = 150$; $S_3 = 150$; $X_2 = 0.7$; $S_{21} = 500$; $S_{22} = 500$; $TP^* = 3,500$; $TP_L = 2,000$; $RR^* = 1,000$; $RR_L = 800$; $NDI^* = 60$; $NDI_U = 100$; $MINTPS = 0.25$

Upon application of the above data to the model using LINGO (v4), the following optimal solution is obtained:

$Z_{12} = 15$; $Z_{13} = 20$; $Z_{21} = 21$; $Z_{22} = 17$; $Z_{31} = 5$; $Z_{32} = 18$; $Z_{33} = 18$; $N_{111} = 7$; $N_{211} = 9$; $A_{121} = 15$; $A_{131} = 20$; $A_{212} = 21$; $A_{222} = 17$; $A_{311} = 5$; $A_{312} = 9$; $A_{321} = 18$; $A_{331} = 17$; $A_{332} = 20$; $B_{111} = 13$; $B_{112} = 15$; $B_{311} = 25$; $B_{312} = 6$; $B_{221} = 7$; $B_{222} = 22$; $B_{322} = 14$; $Y_1 = 1$; $Y_2 = 1$; $TP = 3,500$; $RR = 1,000$; $NDI = 77$; $\mu_1 = 1$; $\mu_2 = 1$; $\mu_3 = 0.59$

It should be noted that the achievements of TP and RR (μ_1 and μ_2) are at their maximum, whereas that of NDI is not (μ_3). In addition, both production facilities are chosen for the network design. Also, 86% of the net demand is satisfied by remanufactured products and the remaining 14% by newly manufactured products.

8.7   Conclusions

In this chapter, five models are presented for achieving transportation of the right quantities of products (used, remanufactured, and new) across a reverse or closed-loop supply chain while satisfying certain constraints. The first and second models are for a reverse supply chain and use linear integer programming and linear physical programming, respectively. The third, fourth, and fifth models are for a closed-loop supply chain and employ goal programming, linear physical programming, and fuzzy goal programming, respectively.

References

1. Fleischmann, M. 2001. *Quantitative models for reverse logistics: Lecture notes in economics and mathematical systems.* Berlin: Springer-Verlag.
2. Messac, A., Gupta, S. M., and Akbulut, B. 1996. Linear physical programming: A new approach to multiple objective optimization. *Transactions of Operational Research* 8:39–59.

9

Evaluation of Marketing Strategies

9.1 The Issue

The success of a reverse/closed-loop supply chain program is heavily dependent on the level of public participation (in the program), which in turn is shouldered by the marketing strategy of that program. Hence, evaluating the marketing strategy of a reverse/closed-loop supply chain program is equivalent to evaluating how well the strategy is driving the public to participate in the program. Studies (for example, see [1] and [2]) are conducted in numerous cities around the world in order to assess the level of participation of the public in the respective reverse/closed-loop supply chain programs. The officials of each program painstakingly approach many homes in the respective city with questions regarding how convenient the program is to the public and how the program can be improved. Although the drivers for governments and companies to implement these programs and evaluate the programs' marketing strategies are environmental consciousness and profitability, respectively, the drivers for the public to participate in the programs are numerous and often conflicting with each other (for example, the more regularly a reverse/closed-loop supply chain program offers to collect used products from consumers, the higher the taxes the consumers will have to pay; high regularity of collection and low tax levied on the consumers are conflicting drivers here).

The various scenarios for evaluating the marketing strategy of a reverse/ closed-loop supply chain program could differ as follows:

1. The program could be exclusively for reverse supply chain operations, i.e., absence of a closed loop.
2. The program could be for a closed-loop supply chain and the decision maker could be uninterested in considering interdependencies among evaluation criteria.
3. The program could be for a closed-loop supply chain and the decision maker could be interested in considering interdependencies among evaluation criteria.

In this chapter, various models to address the above scenarios are presented. The first model addresses scenario 1 and employs fuzzy logic and technique for order preference by similarity to ideal solution (TOPSIS). The second model is for scenario 2 and employs fuzzy logic, quality function deployment (QFD), and method of total preferences. The third model is for scenario 3 and uses fuzzy logic, extent analysis method, and analytic network process. This chapter is organized as follows: Section 9.2 presents the first model, section 9.3 shows the second model, section 9.4 presents the third model, and section 9.5 gives some conclusions.

9.2   First Model (Fuzzy Logic and TOPSIS)

In section 9.2.1, a list of drivers for the public to participate in a reverse supply chain program are identified. Then, in section 9.2.2, using a numerical example, fuzzy logic and TOPSIS are employed to evaluate the marketing strategy of a reverse supply chain program with respect to the drivers identified in section 9.2.1.

9.2.1   Drivers of Public Participation

The following is a fairly exhaustive list of self-explanatory drivers for the public to participate in a reverse supply chain:

1. Knowledge of the drivers of implementation of the reverse supply chain program (KD)
2. Awareness of the reverse supply chain program being implemented (AR)
3. Simplicity of the reverse supply chain program (SR)
4. Convenience for disposal of used products at collection centers (CD)
5. Incentives for disposal of used products (ID)
6. Effectiveness of collection methods (EC)
7. Information supplied about used products being collected (IU)
8. Regularity of collection of used products (RC)
9. Design of special methods for abusers of the reverse supply chain program (AB)
10. Good locations of demand centers where reprocessed goods are sold (LR)
11. Incentives to buyers of reprocessed goods (IB)
12. Cooperation of the program organizers with the local government (CL)

9.2.2 Methodology

Suppose there are three representatives of a community to weigh the drivers of public participation, depending on which driver greatly motivates them to participate, which driver is not so important for them, and so on. Because it is difficult for them to assign numerical weights, they give linguistic weights like "very high," "low," and "medium." Table 9.1 illustrates the linguistic weights. Using fuzzy logic, these linguistic weights are converted into triangular fuzzy numbers (TFNs; table 9.2 shows one of the many ways for such a conversion) and then averaged to form another TFN called the average weight. For example, the average weight of the driver *ID* (see tables 9.1 and 9.2) is

$$\frac{Low + High + Low}{3}$$

TABLE 9.1

Linguistic weights of drivers of public participation (first model)

Driver	Rep. 1	Rep. 2	Rep. 3
KD	Low	Medium	High
AR	Medium	High	Very high
SR	Low	Low	Very high
CD	Very high	Very high	Medium
ID	Low	High	Low
EC	High	Medium	Very high
IU	Low	Low	High
RC	Medium	Low	Low
AB	Medium	Low	Medium
LR	Very high	High	High
IB	High	High	Medium
CL	Medium	High	Low

TABLE 9.2

Conversion table for weights of drivers (first model)

Linguistic weight	TFN
Very high	(0.7, 0.9, 1.0)
High	(0.5, 0.7, 0.9)
Medium	(0.3, 0.5, 0.7)
Low	(0.1, 0.3, 0.5)
Very low	(0.0, 0.1, 0.3)

which is

$$\left(\frac{0.1+0.5+0.1}{3}, \frac{0.3+0.7+0.3}{3}, \frac{0.5+0.9+0.5}{3} \right) = (0.23, 0.43, 0.63)$$

(see table 9.3).

The sum of the average weights of all the drivers is calculated using equation (4.4) as (4.27, 6.33, 8.93). The ratio of the average weight of each driver to the sum of the average weights of all the drivers gives the corresponding normalized weight. For example, the normalized weight of *ID* is

$$\left(\frac{(0.23, 0.43, 0.63)}{(4.27, 6.33, 8.93)} \right)$$

which is simplified using equation (4.7) as (0.03, 0.06, 0.15) (see table 9.4).

Suppose that one is interested in evaluating marketing strategies of two different reverse supply chain programs. Now that the weights (normalized) of the drivers of public participation are ready, the two marketing strategies are linguistically rated by the three representatives with respect to each driver. Table 9.5 is used for conversion of linguistic ratings into TFNs. Assuming (for arithmetic simplicity) that the representatives come to a consensus about the rating of each marketing strategy with respect to each driver, the decision matrix shown in table 9.6 (S1 and S2 are the marketing strategies) is formed. For example, the rating (TFN) of marketing strategy S2 with respect to driver *ID* is (5, 7, 10) (see table 9.6) because the representatives unanimously rate it as good with respect to *ID* (the TFN for the linguistic rating good is (5, 7, 10) (see table 9.5)).

TABLE 9.3

Average weights of drivers of public participation (first model)

Driver	Average weight
KD	(0.3, 0.5, 0.7)
AR	(0.5, 0.7, 0.87)
SR	(0.3, 0.5, 0.67)
CD	(0.57, 0.77, 0.9)
ID	(0.23, 0.43, 0.63)
EC	(0.5, 0.7, 0.87)
IU	(0.23, 0.43, 0.63)
RC	(0.17, 0.37, 0.57)
AB	(0.17, 0.43, 0.63)
LR	(0.57, 0.77, 0.93)
IB	(0.43, 0.63, 0.83)
CL	(0.3, 0.5, 0.7)
Sum	(4.27, 6.73, 8.93)

TABLE 9.4

Normalized weights of drivers of public participation (first model)

Driver	Normalized weight
KD	(0.03, 0.07, 0.16)
AR	(0.06, 0.10, 0.20)
SR	(0.03, 0.07, 0.16)
CD	(0.06, 0.11, 0.21)
ID	(0.03, 0.06, 0.15)
EC	(0.07, 0.10, 0.20)
IU	(0.03, 0.06, 0.15)
RC	(0.02, 0.05, 0.13)
AB	(0.02, 0.06, 0.15)
LR	(0.06, 0.11, 0.22)
IB	(0.05, 0.09, 0.20)
CL	(0.03, 0.07, 0.16)

TABLE 9.5

Conversion for ratings of marketing strategies (first model)

Linguistic rating	TFN
Very good	(7, 10, 10)
Good	(5, 7, 10)
Fair	(2, 5, 8)
Poor	(0, 3, 5)
Very poor	(0, 0, 3)

TABLE 9.6

Decision matrix (first model)

Driver	S1	S2
KD	(7, 10, 10)	(0, 0, 3)
AR	(2, 5, 8)	(7, 10, 10)
SR	(2, 5, 8)	(2, 5, 8)
CD	(0, 0, 3)	(7, 10, 10)
ID	(7, 10, 10)	(5, 7, 10)
EC	(5, 7, 10)	(2, 5, 8)
IU	(0, 3, 5)	(2, 5, 8)
RC	(0, 0, 3)	(7, 10, 10)
AB	(2, 5, 8)	(5, 7, 10)
LR	(5, 7, 10)	(5, 7, 10)
IB	(0, 3, 5)	(7, 10, 10)
CL	(0, 0, 3)	(5, 7, 10)

Now one is ready to perform the six steps in TOPSIS.

Step 1: *Construct the normalized decision matrix.* Table 9.7 shows the normalized decision matrix formed by applying equation (4.49) on each element of table 9.5. For example, the normalized fuzzy rating of strategy S2 with respect to driver *ID* (see table 9.7) is calculated using equation (4.49) as follows:

$$r_{52} = \frac{(5,7,10)}{\sqrt{(7,10,10)^2+(5,7,10)^2}} = (0.35,0.57,1.16)$$

Note that equations (4.4), (4.6), and (4.7) are used to perform the basic operations in the calculation of r_{52} above.

Step 2: *Construct the weighted normalized decision matrix.* Table 9.8 shows the weighted normalized decision matrix. This is constructed using the normalized weights of the drivers listed in table 9.6 and the normalized decision matrix in table 9.7. For example, the weighted normalized fuzzy rating of strategy S2 with respect to driver *ID*, i.e., (0.01, 0.04, 0.17) (see table 9.8), is calculated by multiplying the normalized weight of *ID*, i.e., (0.03, 0.07, 0.15) (see table 9.6), with the normalized fuzzy rating of S2 with respect to *ID*, i.e., (0.35, 0.57, 1.16) (see table 9.7). Equation (4.6) is used for the multiplication.

Step 3: *Determine the ideal and negative-ideal solutions.* Each row in the decision matrix shown in table 9.8 has a maximum rating and a minimum rating. They are the ideal and negative-ideal solutions, respectively, for the corresponding driver. For arithmetic simplicity, it is assumed here that the rating with the highest most promising

TABLE 9.7

Normalized decision matrix (first model)

Driver	S1	S2
KD	(0.67, 1.00, 1.43)	(0.00, 0.00, 0.43)
AR	(0.16, 0.45, 1.10)	(0.55, 0.89, 1.37)
SR	(0.18, 0.71, 2.83)	(0.18, 0.71, 2.83)
CD	(0.00, 0.00, 0.43)	(0.67, 1.00, 1.43)
ID	(0.50, 0.82, 1.16)	(0.35, 0.57, 1.16)
EC	(0.39, 0.81, 1.86)	(0.16, 0.58, 1.49)
IU	(0.00, 0.51, 2.50)	(0.21, 0.86, 4.00)
RC	(0.00, 0.00, 0.43)	(0.67, 1.00, 1.43)
AB	(0.16, 0.58, 1.49)	(0.39, 0.81, 1.86)
LR	(0.35, 0.71, 1.41)	(0.35, 0.71, 1.41)
IB	(0.00, 0.29, 0.71)	(0.63, 0.96, 1.43)
CL	(0.00, 0.00, 0.60)	(0.48, 1.00, 2.00)

TABLE 9.8

Weighted normalized decision matrix (first model)

Driver	S1	S2
KD	(0.02, 0.07, 0.23)	(0.00, 0.00, 0.07)
AR	(0.01, 0.05, 0.22)	(0.03, 0.09, 0.28)
SR	(0.01, 0.05, 0.44)	(0.01, 0.05, 0.44)
CD	(0.00, 0.00, 0.09)	(0.04, 0.11, 0.30)
ID	(0.01, 0.05, 0.17)	(0.01, 0.04, 0.17)
EC	(0.03, 0.08, 0.38)	(0.01, 0.06, 0.30)
IU	(0.00, 0.03, 0.37)	(0.01, 0.06, 0.59)
RC	(0.00, 0.00, 0.06)	(0.01, 0.05, 0.19)
AB	(0.00, 0.04, 0.22)	(0.01, 0.05, 0.28)
LR	(0.02, 0.08, 0.31)	(0.02, 0.08, 0.31)
IB	(0.00, 0.03, 0.14)	(0.03, 0.09, 0.28)
CL	(0.00, 0.00, 0.10)	(0.02, 0.07, 0.33)

quantity (second parameter in the TFN) is the maximum, and the rating with the lowest most promising quantity is the minimum. For example (see table 9.8), with respect to driver *ID*, the maximum rating is (0.01, 0.05, 0.17) and the minimum rating is (0.01, 0.04, 0.17). This is because, in the row for that driver, (0.01, 0.05, 0.17) is the TFN with the highest second parameter and (0.01, 0.04, 0.17) is the TFN with the lowest second parameter.

Step 4: *Calculate the separation distances.* The separation distances (see table 9.9) for each marketing strategy are calculated using equations (4.52) and (4.53). For example, the positive separation distance for strategy S2 (see table 9.9) is calculated using the weighted normalized fuzzy ratings of S2 (see table 9.8) and the ideal solution (obtained in step 3) for each driver of public participation.

Step 5: *Calculate the relative closeness coefficient.* Using equation (4.54), the relative closeness coefficient is calculated for each marketing strategy (see table 9.10). For example, the relative closeness coefficient (i.e., 0.692) for strategy S2 (see table 9.10) is the ratio of the strategy's negative separation distance (i.e., 0.215) to the sum (i.e., 0.215 + 0.097 = 0.312) of its negative and positive separation distances (see table 9.9).

Step 6: *Rank the preference order.* Because the relative closeness coefficient of strategy S2 (i.e., 0.692) is much higher than that of strategy S1 (i.e., 0.308) (see table 9.10), it is evident that S2 is much better than S1.

TABLE 9.9

Separation distances of marketing strategies (first model)

Marketing strategy	Positive distance S*	Negative distance S–
S1	0.215	0.097
S2	0.096	0.215

TABLE 9.10

Relative closeness coefficients of marketing strategies (first model)

Marketing strategy	Relative closeness coefficient
S1	0.311
S2	0.692

9.3   Second Model (Fuzzy Logic, Quality Function Deployment, and Method of Total Preferences)

In this section, a model to evaluate the marketing strategy planned by a closed-loop supply chain program is presented. Fuzzy logic, quality function deployment (QFD), and method of total preferences are employed in the approach. Section 9.3.1 presents the performance aspects and their enablers considered in the use of QFD, and section 9.3.2 gives a numerical example illustrating the model's application.

9.3.1   Performance Aspects and Enablers

The following performance aspects of the marketing strategies are considered in the employment of quality function deployment:

1. *Program simplicity* (*PS*) (reflects the relative ease with which the public can participate in the program)
2. *Manufacturing practices* (*MP*) (reflects the production facility's green image, innovation, and improvement capability compared to its peers)
3. *Government issues* (*GI*) (reflects how stringent the local government regulations are)
4. *Incentives* (reflects the incentives offered by the collection centers and the local government to the public for participating in the program)
5. *Public knowledge* (*PK*) (reflects the awareness among the public about the program being implemented)

The following enablers are considered for the above-listed performance aspects:

1. *Program simplicity (PS)*: Strategic location of collection centers (*SL*), effectiveness of collection methods (*ECM*)

2. *Manufacturing practices (MP)*: Green competency (*GC*), innovation, and improvement capability (*I&I*)

3. *Government issues (GI)*: Regulations (*Reg.*), company's level of cooperation with local governments (*Co-op*)

4. *Incentives*: Disposal incentives by the local government and the collection centers (*DI*), incentives for buying remanufactured goods (*BI*)

5. *Public knowledge (PK)*: Good advertisement of the program being implemented (*AD*), socioeconomic status of the society where the program is implemented (*SES*)

For the convenience of the reader, the performance aspects and the respective enablers are shown in a four-level hierarchy (see figure 9.1). The first level represents the goal (evaluating the marketing strategies), the second level contains the performance aspects, the third level contains the enablers of the performance aspects considered in the second level, and the fourth level contains the candidate marketing strategies (A and B) to be evaluated.

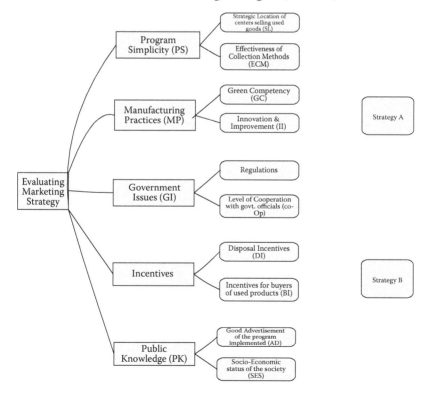

FIGURE 9.1
Performance aspects and enablers of marketing strategies (second model).

9.3.2   Numerical Example

Suppose that there are three experts (E_1, E_2, and E_3) for giving linguistic values to R_{ij} (relationship score for performance aspect i and enabler j) and d_i (importance value of performance aspect i relative to the other performance aspects) data. Table 9.11 shows the linguistic scale for R_{ij} and d_i.

Tables 9.12–9.16 show the linguistic relationship scores (R_{ij}) as given by three experts to the enablers of *PS*, *MP*, *GI*, *Incentives*, and *PK*, respectively.

Table 9.17 gives the linguistic importance values (d_i) of *PS*, *MP*, *GI*, *Incentives*, and *PK* as given by the three experts.

These linguistic relationship scores are converted into TFNs using the conversion scale shown in table 9.11. The TFNs are averaged before defuzzifying. For example, consider *SL*'s linguistic relationship scores given by the three experts (see table 9.12). The average relationship score for *SL*, calculated using equations (4.4) and (4.7), is

$$\left(\frac{7.5+5+7.5}{3}, \frac{10+7.5+10}{3}, \frac{10+10+10}{3}\right) = (6.67, 9.17, 10)$$

The defuzzified relationship score for *SL*, calculated using equation (4.8), is 8.61. The average relationship scores and the corresponding defuzzified values for *PS*, *MP*, *GI*, *Incentives*, and *PK* are calculated as (6.67, 9.17, 10) (defuzzified value = 8.61), (4.17, 6.67, 9.17) (defuzzified value = 6.67), (5.83, 8.33, 9.17) (defuzzified value = 7.78), (5.83, 8.33, 10) (defuzzified value = 8.05), and (3.33, 6.83, 8.33) (defuzzified value = 5.83), respectively.

The ATIRs and RTIRs of the enablers are then calculated using equations (4.30) and (4.31), respectively (see table 9.18).

Table 9.19 shows the scale for converting linguistic WA_{nj} (degree to which alternative n can deliver enabler j) values given by the three experts into TFNs (for arithmetic simplicity, we assume consensus among the experts).

The WA_{nj} linguistic value and the corresponding defuzzified TFN (calculated using equation (4.8)) for each alternative are shown in table 9.20.

Using equation (4.32), the *TUP* value for each alternate marketing strategy is calculated. For example, *TUP* for marketing strategy A is calculated using the *RTIR* values from table 9.18 and the defuzzified WA_{nj} values from table 9.20 as (0.13)×(10) + (0.095)×(7.33) + (0.087)×(10) + (0.077)×(5) + (0.121)×(5) + (0.117)×(7.33) + (0.089)×(7.33) + (0.125)×(5) + (0.0821)×(3) + (0.07)×(5) = 6.64.

Finally, using equation (4.33), the *NTUP* value for each alternative marketing strategy is calculated. For example, *NTUP* for marketing strategy A is calculated as

$$\frac{6.64}{6.64+6.62} = 0.5007$$

It is evident that marketing strategy A has more potential than marketing strategy B, and hence the decision maker would choose strategy A.

TABLE 9.11

Linguistic weight conversion table for R_{ij} and d_i (second model)

Linguistic rating	TFN
Very strong (VS)	(7.5, 10, 10)
Strong (S)	(5, 7.5, 10)
Medium (M)	(2.5, 5, 7.5)
Weak (W)	(0, 2.5, 5)
Very weak (VW)	(0, 0, 2.5)

TABLE 9.12

Linguistic relationship scores of PS and its enablers (second model)

	E1	E2	E3
SL	VS	S	VS
ECM	S	M	S

TABLE 9.13

Linguistic relationship scores of MP and its enablers (second model)

	E1	E2	E3
GC	VS	VS	S
I&I	S	S	VS

TABLE 9.14

Linguistic relationship scores of GI and its enablers (second model)

	E1	E2	E3
Reg.	S	VS	M
Co-op	VS	M	M

TABLE 9.15

Linguistic relationship scores of incentives and its enablers (second model)

	E1	E2	E3
DI	VS	S	VS
BI	S	M	S

TABLE 9.16

Linguistic relationship scores of PK and its enablers (second model)

	E1	E2	E3
AD	VS	M	VS
SES	M	S	S

TABLE 9.17

Linguistic relationship scores of performance aspects (second model)

	E1	E2	E3
PS	S	VS	VS
MP	S	S	M
GI	VS	VS	M
Incentives	S	VS	S
PK	S	M	M

TABLE 9.18

ATIRs and RTIRs of enablers (second model)

Enabler	ATIR	RTIR
SL	74.15	0.13
ECM	52.62	0.10
GC	48.15	0.09
I&I	42.59	0.08
Reg.	66.98	0.12
Co-op	64.81	0.12
DI	49.23	0.09
BI	69.37	0.13
AD	45.37	0.08
SES	38.89	0.07

TABLE 9.19

Conversion table for linguistic WA_{nj} (second model)

Linguistic rating	TFN
Very good (VG)	(7, 10, 10)
Good (G)	(5, 7, 10)
Fair (F)	(2, 5, 8)
Poor (P)	(1, 3, 5)
Very poor (VP)	(0, 0, 3)

9.4 Third Model (Fuzzy Logic, Extent Analysis Method, and Analytic Network Process)

In this section, a model to evaluate the marketing strategy planned by a closed-loop supply chain program is presented. Fuzzy logic and analytic network process are employed in the approach. The difference between this model and the second one (see section 9.3) is that interdependencies among criteria are considered here (unlike in the second model). Section 9.4.1 pres-

TABLE 9.20

Linguistic and corresponding defuzzified *WA* values of marketing strategies (second model)

Enabler	A		B	
SL	VG	10	G	7.33
ECM	G	7.33	F	5
GC	VG	10	VG	10
I&I	F	5	VG	10
Reg	F	5	G	7.33
Co-op	G	7.33	G	7.33
DI	G	7.33	F	5
BI	F	5	P	3
AD	P	3	P	3
SES	F	5	VG	10

ents the main criteria and subcriteria for evaluation, and section 9.4.2 gives a numerical example illustrating the model's application.

9.4.1 Main Criteria and Subcriteria

The main criteria and subcriteria are shown in a four-level hierarchy (see figure 9.2; differs slightly from figure 9.1). For a description of these criteria, see section 9.3.1 (performance aspects and enablers). The first level represents the goal (evaluating the marketing strategies), the second level contains the main criteria, the third level contains the subcriteria of the main criteria considered in the second level, and the fourth level contains the candidate marketing strategies (C and D) to be evaluated.

9.4.2 Numerical Example

Suppose that there are three experts to carry out pair-wise comparisons among the main criteria and subcriteria using linguistic weights (high, medium, low, etc.). These linguistic weights are converted into TFNs using table 9.21.

Table 9.22 illustrates the comparative linguistic weights (H = high, M = medium, L = low) given to the main criteria. Using fuzzy logic, these linguistic weights are converted into TFNs (using table 9.21) and then averaged to form another TFN called the average weight. For example, the average weight of "simplicity" criteria with respect to government issues (*GI*) is

$$\frac{H + H + M}{3}$$

which is

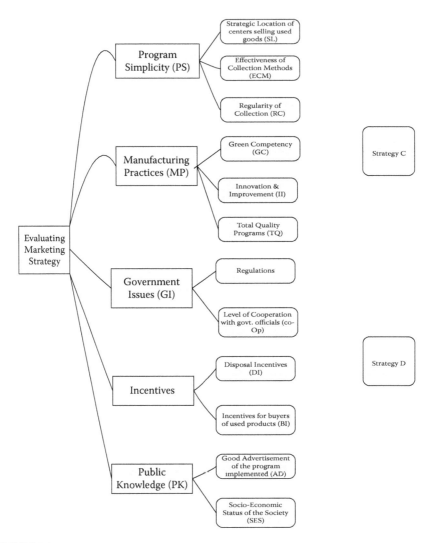

FIGURE 9.2
Main criteria and subcriteria (third model).

$$\left(\frac{0.5+0.5+0.3}{3}, \frac{0.7+0.7+0.5}{3}, \frac{0.9+0.9+0.7}{3} \right) = (0.43, 0.63, 0.83)$$

(see table 9.23).

Then the extent analysis method (see equations (4.9)–(4.11)) is applied to the average weights of the main criteria to calculate the fuzzy synthetic extent values for the same. By applying equations (4.12)–(4.14) to the fuzzy synthetic extent values, the weight vectors are obtained for the main criteria, which are then normalized to get the normalized weights shown in table 9.24.

TABLE 9.21

Linguistic weight conversion table for criteria and subcriteria (third model)

Linguistic weight	TFN
Very high (VH)	(0.7, 0.9, 1.0)
High (H)	(0.5, 0.7, 0.9)
Medium (M)	(0.3, 0.5, 0.7)
Low (L)	(0.1, 0.3, 0.5)
Very low (VL)	(0.0, 0.1, 0.3)

TABLE 9.22

Linguistic weights of main criteria (third model)

Criteria	PS	GI	Incentives	MP	PK
PS	(1, 1, 1)	(H, H, M)	(M, L, M)	(L, VL, M)	(L, VL, L)
GI	1/(H, H, M)	(1, 1, 1)	(VH, H, H)	(H, M, H)	(M, H, M)
Incentives	1/(M, L, M)	1/(VH, H, H)	(1, 1, 1)	(H, M, H)	(M, L, M)
MP	1/(L, VL, M)	1/(H, M, H)	1/(H, M, H)	(1, 1, 1)	(H, M, H)
PK	1/(L, VL, L)	1/(M, H, M)	1/(M, L, M)	1/(H, M, H)	(1, 1, 1)

TABLE 9.23

Average weights of main criteria (third model)

Criteria	PS	GI	Incentives	MP	PK
PS	(1, 1, 1)	(0.43, 0.63, 0.83)	(0.23, 0.43, 0.63)	(0.14, 0.3, 0.5)	(0.07, 0.23, 0.43)
GI	(1.20, 1.58, 2.32)	(1, 1, 1)	(0.57, 0.77, 0.63)	(0.43, 0.63, 0.83)	(0.37, 0.57, 0.77)
Incentives	(1.58, 2.32, 4.34)	(1.58, 1.29, 1.75)	(1, 1, 1)	(0.43, 0.63, 0.83)	(0.23, 0.43, 0.63)
MP	(2, 3.33, 7.14)	(1.2, 1.58, 2.32)	(1.2, 1.58, 2.32)	(1, 1, 1)	(0.43, 0.63, 0.83)
PK	(2.32, 4.34, 4.28)	(1.29, 0.07, 2.7)	(1.58, 2.32, 4.34)	(1.2, 1.58, 2.32)	(1, 1, 1)
Sum	(8.11, 12.59, 29.1)	(5.52, 4.58, 8.61)	(4.59, 6.11, 8.93)	(3.2, 4.14, 5.58)	(2.1, 2.86, 3.66)

TABLE 9.24

Normalized weights of main criteria (third model)

Criteria	Weight
PS	0.015
GI	0.131
Incentives	0.217
MP	0.304
PK	0.331

In a similar manner, the normalized weight of each subcriterion with respect to its main criteria is obtained (see table 9.25).

Table 9.26 shows the matrix of interdependencies obtained after carrying out pair-wise comparisons among the subcriteria and the steps involved in the extent analysis method.

The super matrix M is made to converge to obtain a long-term stable set of weights. For convergence, M must be made column stochastic, which is done by raising M to the power of 2^{k+1}, where k is an arbitrarily large number; in the example here, $k = 59$. Table 9.27 shows the converged super matrix.

TABLE 9.25

Normalized weights of subcriteria with respect to main criteria (third model)

Subcriteria	Weight
SL	0.37
ECM	0.33
RC	0.30
Regulations	0.12
Co-op	0.88
GC	0.80
I&I	0.20
TQ	0.06
DI	0.29
BI	0.65
AD	0.89
SES	0.11

TABLE 9.26

Matrix of interdependencies (third model)

	SL	ECM	RC	Reg.	Co-op	GC	I&I	TQ	DI	BI	AD	SES
SL	0	0.2	0.08	0	0	0	0	0	0	0	0	0
ECM	0.11	0	0.92	0	0	0	0	0	0	0	0	0
RC	0.889	0.8	0	0	0	0	0	0	0	0	0	0
Regulations	0	0	0	0	0.5	0	0	0	0	0	0	0
Co-op	0	0	0	0.5	0	0	0	0	0	0	0	0
GC	0	0	0	0	0	0	0.51	0.259	0	0	0	0
I&I	0	0	0	0	0	0.11	0	0.74	0	0	0	0
TQ	0	0	0	0	0	0.88	0.48	0	0	0	0	0
DI	0	0	0	0	0	0	0	0	0	0.5	0	0
BI	0	0	0	0	0	0	0	0	0.5	0	0	0
AD	0	0	0	0	0	0	0	0	0	0	0	0.5
SES	0	0	0	0	0	0	0	0	0	0	0.5	0

TABLE 9.27

Converged super matrix (third model)

SL	0.121
ECM	0.487
RC	0.45
Regulations	0
Co-op	0
GC	0.271
I&I	0.326
TQ	0.401
DI	0
BI	0
AD	0
SES	0

Table 9.28 shows the linguistic ratings (and their corresponding TFNs) that are used to evaluate the marketing strategies with respect to the subcriteria.

Relative ratings of the marketing strategies (C and D) are obtained by carrying out pair-wise comparisons among the marketing strategies with respect to the subcriteria and applying the steps involved in the extent analysis method. In the example here, there are seventeen subcriteria that lead to seventeen such pair-wise comparison matrices. Table 9.29 shows the relative ratings of the marketing strategies.

Using equation (4.2), the desirability indices (*DI*) for each marketing strategy are calculated (see table 9.30).

The overall performance index for each marketing strategy is calculated by multiplying the desirability indices (table 9.30) of each marketing strategy by the weights of the criteria (table 9.24) and summing up over all the criteria and normalizing those indices. Table 9.31 shows the overall weighted indices for the two marketing strategies.

Strategy D's overall weighted index is greater; hence, the decision maker would choose marketing strategy D.

TABLE 9.28

Linguistic weight conversion table for ratings of marketing strategies (third model)

Linguistic weight	TFN
Very good (VG)	(7, 10, 10)
Good (G)	(5, 7, 10)
Fair (F)	(2, 5, 8)
Poor (P)	(1, 3, 5)
Very poor (VP)	(1, 1, 3)

TABLE 9.29

Relative ratings of marketing strategies with respect to subcriteria (third model)

	C	D
SL	0.50	0.50
ECM	0.48	0.52
RC	0.50	0.50
Regulations	0.50	0.50
Co-op	0.48	0.52
GC	0.48	0.52
I&I	0.48	0.52
TQ	0.49	0.51
DI	0.48	0.52
BI	0.50	0.50
AD	0.49	0.51
SES	0.48	0.52

TABLE 9.30

Desirability indices (third model)

Criterion	S1	S2
PS	0.16	0.16
GI	0.00	0.00
Incentives	0.18	0.19
MP	0.00	0.00
PK	0.00	0.00

TABLE 9.31

Overall weighted indices for marketing strategies (third model)

Marketing strategy	Overall weighted index
C	0.49
D	0.51

9.5 Conclusions

In this chapter, three models to evaluate marketing strategies for a reverse/closed supply chain program are presented. The first model employs fuzzy logic and technique for order preference by similarity to ideal solution (TOPSIS). The second model employs fuzzy logic, quality function deployment (QFD), and method of total preferences. Finally, the third model uses fuzzy logic, extent analysis method, and analytic network process. Whereas the first model is for a reverse supply chain program, the second and third models are for a closed-loop supply chain program. The difference between the second and third models is that interdependencies among criteria are considered in the third model, but not in the second one.

References

1. McDonald, S., and Ball, R. 1998. Public participation in plastics recycling schemes. *Resources, Conservation, and Recycling* 22:123–41.
2. Williams, I. D., and Kelly, J. 2003. Green waste collection and the public's recycling behavior in the Borough of Wyre England. *Resources, Conservation, and Recycling* 38:139–59.

10

Evaluation of Production Facilities

10.1 The Issue

In addition to selecting efficient collection centers (see chapter 6), strategic planning of a closed-loop supply chain involves selecting efficient production facilities where not only new products are produced but also used products are reprocessed. The various scenarios for evaluating production facilities for efficiency could differ as follows:

1. The decision maker could desire to structure the problem using a simple hierarchical model, wherein interactions among evaluation criteria could be ignored.
2. Interactions among evaluation criteria could not be ignored, which in turn leads to a more complex problem structure.
3. Interactions among evaluation criteria could not be ignored, and the decision maker could desire to see how close the rating of a candidate production facility is to the "ideal" solution.

In this chapter, three models to address the above scenarios are presented. These models evaluate production facilities in terms of both environmental consciousness and potentiality. The first model addresses scenario 1 and employs fuzzy logic and technique for order preference by similarity to ideal solution (TOPSIS). The second model is for scenario 2 and employs fuzzy logic, extent analysis method, and analytic network process. The third model is for scenario 3 and employs fuzzy multicriteria analysis method.

This chapter is organized as follows: section 10.2 presents the first model, section 10.3 shows the second model, section 10.4 presents the third model, and section 10.4 gives some conclusions.

10.2   First Model (Fuzzy Logic and TOPSIS)

The problem of evaluation of production facilities for designing a closed-loop supply chain is framed as a four-level hierarchy. The first level in the hierarchy contains the objective, i.e., evaluation of the efficiency of each candidate production facility; the second level contains the main criteria for evaluation; the third level contains the subcriteria under each main criterion; and the fourth (last) level contains the candidate production facilities. Figure 10.1 illustrates these hierarchy levels.

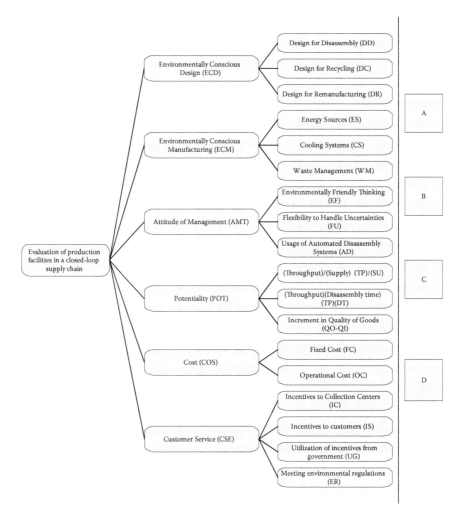

FIGURE 10.1
Levels of hierarchy.

Section 10.2.1 gives brief descriptions of the main and subcriteria for evaluation, and section 10.2.2 illustrates the model using a numerical example.

10.2.1 Evaluation Criteria

The following are brief descriptions of the main criteria on the second level in the hierarchy and the corresponding subcriteria on the third level (see [1] for detailed descriptions of some of the criteria).

10.2.1.1 *Environmentally Conscious Design* (ECD)

ECD is concerned with designing products with certain environmental considerations. The following subcriteria fall under *ECD*.

10.2.1.1.1 *Design for Disassembly (DD)*

Disassembly is used in both recycling and remanufacturing to increase the recovery rate by allowing selective separation of parts and materials. Thus, *DD* initiatives lead to the correct identification of design specifications to minimize the complexity of the structure of the product by minimizing the number of parts, increasing the use of common materials, and choosing a fastener and joint types that are easily removable.

10.2.1.1.2 *Design for Recycling (DC)*

DC suggests making better choices for material selection such that the processes of material selection and material recovery become more efficient. Some important characteristics of *DC* are long product life with the minimized use of raw materials (source reduction), more adaptable materials for multiproduct applications, and fewer components within a given material in an engineered system.

10.2.1.1.3 *Design for Remanufacturing (DR)*

DR suggests the use of reusable parts and packaging in the design of new products for source reduction.

10.2.1.2 *Environmentally Conscious Manufacturing* (ECM)

In addition to environmentally friendly product designs, issues involving manufacturing must be addressed to have a complete concept of environmentally conscious production. These issues (subcriteria on the third level in the hierarchy) are the following:

- Selecting low-pollution energy sources for manufacturing (*ES*)
- Designing cooling systems such that the coolant can be reused and the heat collected by it can be utilized as an energy source (*CS*)
- Monitoring waste generation as a result of manufacturing (*WM*)

10.2.1.3   Attitude of Management *(AMT)*

The attitude of the decision makers (managers) in the facility matters a great deal when it comes to implementing the above practices (*ECD* and *ECM*). All the managers in the facility must have the following credentials (subcriteria on the third level in the hierarchy):

- Environmentally friendly thoughts (*EF*)
- Flexibility to handle uncertainties in the supply and quality of used products (*FU*)
- Readiness to usage of automated disassembly systems (*AD*) to avoid high lead time, expensive labor use, and possible human exposure to hazardous by-products

10.2.1.4   Potentiality *(POT)*

A candidate production facility is also evaluated in terms of its potentiality to efficiently reprocess the incoming used products. The following factors (subcriteria on the third level in the hierarchy) serve as potentiality measures (these factors are explained in detail in section 7.2.1):

- Throughput (*TP*)/supply (*SU*)
- Throughput (*TP*)×disassembly time (*DT*)
- Quality of reprocessed products (*QO*)–quality of used products (*QI*)

10.2.1.5   Cost *(COS)*

Cost incurred by a production facility can be divided into the following types (subcriteria on the third level in the hierarchy):

- Fixed cost (*FC*), which is the sum of space cost, machinery cost, personnel cost, etc.
- Operational cost (*OC*), which is the sum of employee salaries, maintenance cost, etc.

10.2.1.6   Customer Service *(CSE)*

Customer service basically gives an idea about how well a production facility is:

- Giving incentives to the collection centers supplying used products (*IC*)
- Giving incentives to the customers buying reprocessed goods (*IS*)
- Utilizing incentives provided by the government (*UG*)
- Meeting environmental regulations established by the government (*ER*)

Note that the term *customer service* is used here because any beneficiary is a customer, be it the government, the collection center, or the actual customer buying reprocessed goods.

10.2.2 Numerical Example

In this example, linguistic (high, medium, etc.) weights are given by three experts to the main criteria (second level in the hierarchy) and the subcriteria (third level in the hierarchy). These linguistic weights are quantified using triangular fuzzy numbers (TFNs) in table 10.1.

Table 10.2 shows the linguistic weights given by the three experts to the main criteria. These are quantified using table 10.1 and then averaged to form another TFN called the average weight. For example, the average weight of the main criterion, *ECD*, is (H + H + M)/3, which is

$$\left(\frac{0.5+0.5+0.3}{3}, \frac{0.7+0.7+0.5}{3}, \frac{0.9+0.9+0.7}{3} \right) = (0.43, 0.63, 0.83)$$

(see table 10.2).

The sum of the average weights of all the main criteria is calculated, using equation (4.4), as (2.29, 3.45, 4.59). The ratio of the average weight of each main

TABLE 10.1

Linguistic weight conversion table for criteria and subcriteria (first model)

Linguistic weight	Triangular fuzzy number
Very high (VH)	(0.7, 0.9, 1.0)
High (H)	(0.5, 0.7, 0.9)
Medium (M)	(0.3, 0.5, 0.7)
Low (L)	(0.1, 0.3, 0.5)
Very low (VL)	(0.0, 0.1, 0.3)

TABLE 10.2

Relative weights of main criteria (first model)

Criterion	Expert E1	Expert E2	Expert E3	Average weight	Relative weight
ECD	H	H	M	(0.43, 0.63, 0.83)	(0.09, 0.18, 0.36)
ECM	VH	H	VH	(0.63, 0.83, 0.97)	(0.14, 0.24, 0.42)
AMT	L	L	VL	(0.07, 0.23, 0.43)	(0.02, 0.07, 0.19)
POT	H	H	H	(0.50, 0.70, 0.90)	(0.11, 0.20, 0.39)
COS	M	H	H	(0.43, 0.63, 0.83)	(0.09, 0.18, 0.36)
CSE	M	L	M	(0.23, 0.43, 0.63)	(0.05, 0.12, 0.28)

criterion to the sum of the average weights of all the main criteria gives the corresponding relative weight. For example, the relative weight of *ECD* is

$$\left(\frac{(0.43, 0.63, 0.83)}{(2.29, 3.45, 4.59)} \right)$$

which is simplified, using equation (4.7), as (0.09, 0.18, 0.36) (see table 10.2).

Similarly, linguistic weights, average weights, and relative weights of subcriteria of *ECD, ECM, AMT, POT, COS*, and *CSE* are calculated and shown in tables 10.3–10.8, respectively.

TABLE 10.3

Relative weights of subcriteria of *ECD* (first model)

Subcriterion	E1	E2	E3	Average weight	Relative weight
DD	H	H	M	(0.43, 0.63, 0.83)	(0.21, 0.42, 0.89)
DC	L	L	VL	(0.07, 0.23, 0.43)	(0.03, 0.15, 0.46)
DR	M	H	H	(0.43, 0.63, 0.83)	(0.21, 0.42, 0.89)

TABLE 10.4

Relative weights of subcriteria of *ECM* (first model)

Subcriterion	E1	E2	E3	Average weight	Relative weight
ES	H	H	H	(0.50, 0.70, 0.90)	(0.21, 0.40, 0.78)
CS	M	H	H	(0.43, 0.63, 0.83)	(0.18, 0.36, 0.72)
WM	M	L	M	(0.23, 0.43, 0.63)	(0.10, 0.24, 0.54)

TABLE 10.5

Relative weights of subcriteria of *AMT* (first model)

Subcriterion	E1	E2	E3	Average weight	Relative weight
EF	L	L	VL	(0.07, 0.23, 0.43)	(0.04, 0.18, 0.59)
FU	M	H	H	(0.43, 0.63, 0.83)	(0.23, 0.49, 1.14)
AD	M	L	M	(0.23, 0.43, 0.63)	(0.12, 0.33, 0.86)

TABLE 10.6

Relative weights of subcriteria of *POT* (first model)

Subcriterion	E1	E2	E3	Average weight	Relative weight
TP/SU	H	H	M	(0.43, 0.63, 0.83)	(0.21, 0.42, 0.89)
TP×DT	L	L	VL	(0.07, 0.23, 0.43)	(0.03, 0.15, 0.46)
QO–QI	M	H	H	(0.43, 0.63, 0.83)	(0.21, 0.42, 0.89)

TABLE 10.7

Relative weights of subcriteria of *COS* (first model)

Subcriterion	E1	E2	E3	Average weight	Relative weight
FC	H	H	M	(0.43, 0.63, 0.83)	(0.34, 0.73, 1.66)
OC	L	L	VL	(0.07, 0.23, 0.43)	(0.06, 0.27, 0.86)

TABLE 10.8

Relative weights of subcriteria of *CSE* (first model)

Subcriterion	E1	E2	E3	Average weight	Relative weight
IC	M	H	H	(0.43, 0.63, 0.83)	(0.17, 0.35, 0.78)
IS	L	L	VL	(0.07, 0.23, 0.43)	(0.03, 0.13, 0.40)
UG	H	H	H	(0.50, 0.70, 0.90)	(0.19, 0.39, 0.84)
ER	L	L	VL	(0.07, 0.23, 0.43)	(0.03, 0.13, 0.40)

Because the weights considered in the TOPSIS method must sum to unity, the weight of each subcriterion on the third level in the hierarchy is multiplied by the weight of its corresponding main criterion on the second level in the hierarchy. The weights of the criteria on the third level in the hierarchy, which are ready for use in the TOPSIS method, are shown in table 10.9. Table 10.10 shows the linguistic ratings (performance measures) and their corresponding TFNs, which are used to evaluate the production facilities with respect to each subcriterion, except *TP/SU, TP×DT, QO–QI, FC,* and *OC*. The reason for this exception is that historical crisp (nonfuzzy) measures of these subcriteria (viz., *TP/SU, TP×DT, QO–QI, FC,* and *OC*) for each production facility can be easily obtained. Table 10.11 shows the crisp measures that are considered for these subcriteria for each production facility.

Now one is ready to perform the six steps in the TOPSIS.

Step 1: *Construct the normalized decision matrix.* Table 10.12 shows the decision matrix whose elements are the fuzzy ratings of the production facilities with respect to each subcriterion, as given by the three experts (for arithmetic simplicity, it is assumed here that the experts give a consensus rating). For example, the rating of facility C with respect to subcriterion *DD* is (1, 3, 5) (see table 10.12) because the experts unanimously rate it as poor with respect to *DD* (the TFN for the linguistic rating "poor" is (1, 3, 5) (see table 10.10)). Also, for consistency in the TOPSIS, the crisp measures of subcriteria *TP/SU, TP×DT, QO–QI, FC,* and *OC* are converted into TFNs, each of whose parameters are all equal. For example, crisp measure 10 of subcriterion *FC* for facility A is converted into the TFN (10, 10, 10) (see table 10.12). Table 10.13 shows the normalized decision matrix formed by applying equation (4.49) to each element of table 10.12. For example, the normalized fuzzy rating of facility C with respect to

TABLE 10.9

Weights of subcriteria for TOPSIS (first model)

Subcriterion	Weight for TOPSIS (relative weight of subcriterion × relative weight of corresponding main criterion)
DD	(0.02, 0.08, 0.32)
DC	(0.00, 0.03, 0.17)
DR	(0.02, 0.08, 0.32)
ES	(0.03, 0.10, 0.33)
CS	(0.03, 0.09, 0.30)
WM	(0.01, 0.06, 0.23)
EF	(0.00, 0.01, 0.11)
FU	(0.00, 0.03, 0.22)
AD	(0.00, 0.02, 0.16)
TP/SU	(0.02, 0.08, 0.35)
TP8 × DT	(0.00, 0.03, 0.18)
QO–QI	(0.02, 0.08, 0.35)
FC	(0.03, 0.13, 0.60)
OC	(0.00, 0.05, 0.31)
IC	(0.01, 0.04, 0.22)
IS	(0.00, 0.02, 0.11)
UG	(0.01, 0.05, 0.24)
ER	(0.00, 0.02, 0.11)

TABLE 10.10

Linguistic rating conversion table for production facilities (first model)

Linguistic rating	Triangular fuzzy number
Very good (VG)	(7, 10, 10)
Good (G)	(5, 7, 10)
Fair (F)	(2, 5, 8)
Poor (P)	(1, 3, 5)
Very poor (VP)	(0, 0, 3)

TABLE 10.11

Crisp measures of subcriteria for evaluation (first model)

Subcriterion	Production facilities			
	A	**B**	**C**	**D**
TP/SU	0.9	0.7	0.9	0.5
TP×DT	25	30	15	40
QO–QI	0.6	0.7	0.3	0.5
FC ($)	100,000	150,000	70,000	200,000
OC ($)	500	300	450	400

TABLE 10.12

Decision matrix for TOPSIS (first model)

Subcriterion	Production facilities			
	A	B	C	D
DD	(7, 10, 10)	(2, 5, 8)	(1, 3, 5)	(5, 7, 10)
DC	(5, 7, 10)	(0, 0, 3)	(5, 7, 10)	(5, 7, 10)
DR	(2, 5, 8)	(7, 10, 10)	(2, 5, 8)	(5, 7, 10)
ES	(1, 3, 5)	(5, 7, 10)	(1, 3, 5)	(1, 3, 5)
CS	(0, 0, 3)	(5, 7, 10)	(1, 3, 5)	(0, 0, 3)
WM	(1, 3, 5)	(0, 0, 3)	(2, 5, 8)	(2, 5, 8)
EF	(1, 3, 5)	(5, 7, 10)	(5, 7, 10)	(7, 10, 10)
FU	(5, 7, 10)	(7, 10, 10)	(2, 5, 8)	(7, 10, 10)
AD	(0, 0, 3)	(7, 10, 10)	(1, 3, 5)	(0, 0, 3)
TP/SU	(0.9, 0.9, 0.9)	(0.7, 0.7, 0.7)	(0.9, 0.9, 0.9)	(0.5, 0.5, 0.5)
TP×DT	(25, 25, 25)	(30, 30, 30)	(15, 15, 15)	(40, 40, 40)
QO–QI	(0.6, 0.6, 0.6)	(0.7, 0.7, 0.7)	(0.3, 0.3, 0.3)	(0.5, 0.5, 0.5)
FC	(10, 10, 10)	(15, 15, 15)	(7, 7, 7)	(20, 20, 20)
OC	(5, 5, 5)	(3, 3, 3)	(4.5, 4.5, 4.5)	(4, 4, 4)
IC	(0, 0, 3)	(0, 0, 3)	(5, 7, 10)	(1, 3, 5)
IS	(1, 3, 5)	(2, 5, 8)	(7, 10, 10)	(5, 7, 10)
UG	(2, 5, 8)	(0, 0, 3)	(1, 3, 5)	(0, 0, 3)
ER	(1, 3, 5)	(5, 7, 10)	(5, 7, 10)	(2, 5, 8)

subcriterion *DD* (see table 10.13) is calculated using equation (4.49) as follows:

$$r_{13} = \frac{(1,3,5)}{\sqrt{(7,10,10)^2 + (2,5,8)^2 + (1,3,5)^2 + (5,7,10)^2}} = (0.06, 0.22, 0.56)$$

Note that equations (4.4) and (4.7) are used to perform the basic operations in the calculation of r_{13} above.

Step 2: *Construct the weighted normalized decision matrix.* Table 10.14 shows the weighted normalized decision matrix. This is constructed using the weights of the subcriteria listed in table 10.9 and the normalized decision matrix in table 10.13. For example, the weighted normalized fuzzy rating of facility C with respect to subcriterion *DD*, i.e., (0, 0.02, 0.18) (see table 10.14), is calculated by multiplying the weight of *DD*, i.e., (0.02, 0.08, 0.32) (see table 10.9), by the normalized fuzzy rating of facility C with respect to *DD*, i.e., (0.06, 0.22, 0.56) (see table 10.13). Equation (4.6) is used for the multiplication.

TABLE 10.13

Normalized decision matrix (first model)

	Production facilities			
Subcriterion	A	B	C	D
DD	(0.41, 0.74, 1.13)	(0.12, 0.37, 0.90)	(0.06, 0.22, 0.56)	(0.29, 0.52, 1.13)
DC	(0.28, 0.58, 1.15)	(0, 0, 0.35)	(0.28, 0.58, 1.15)	(0.28, 0.58, 1.15)
DR	(0.11, 0.35, 0.88)	(0.39, 0.71, 1.1)	(0.11, 0.35, 0.88)	(0.28, 0.50, 1.1)
ES	(0.11, 0.58, 2.89)	(0, 0, 1.73)	(0.11, 0.58, 2.89)	(0.11, 0.58, 2.89)
CS	(0, 0, 0.59)	(0.42, 0.92, 1.96)	(0.08, 0.39, 0.98)	(0, 0, 0.59)
WM	(0.08, 0.39, 1.67)	(0, 0, 1)	(0.16, 0.65, 2.67)	(0.16, 0.65, 2.67)
EF	(0.06, 0.21, 0.5)	(0.28, 0.49, 1)	(0.28, 0.49, 1)	(0.39, 0.7, 1)
FU	(0.26, 0.42, 0.89)	(0.37, 0.6, 0.89)	(0.1, 0.3, 0.71)	(0.37, 0.6, 0.89)
AD	(0, 0, 0.42)	(0.59, 0.96, 1.41)	(0.08, 0.29, 0.71)	(0, 0, 0.42)
TP/SU	(0.59, 0.59, 0.59)	(0.46, 0.46, 0.46)	(0.59, 0.59, 0.59)	(0.33, 0.33, 0.33)
TP×DT	(0.43, 0.43, 0.43)	(0.52, 0.52, 0.52)	(0.26, 0.26, 0.26)	(0.69, 0.69, 0.69)
QO–QI	(0.55, 0.55, 0.55)	(0.64, 0.64, 0.64)	(0.28, 0.28, 0.28)	(0.46, 0.46, 0.46)
FC	(0.36, 0.36, 0.36)	(0.54, 0.54, 0.54)	(0.25, 0.25, 0.25)	(0.72, 0.72, 0.72)
OC	(0.6, 0.6, 0.6)	(0.36, 0.36, 0.36)	(0.54, 0.54, 0.54)	(0.48, 0.48, 0.48)
IC	(0, 0, 0.59)	(0, 0, 0.59)	(0.42, 0.92, 1.96)	(0.08, 0.39, 0.98)
IS	(0.06, 0.22, 0.56)	(0.12, 0.37, 0.90)	(0.41, 0.74, 1.13)	(0.29, 0.52, 1.13)
UG	(0.19, 0.86, 3.58)	(0, 0, 1.34)	(0.1, 0.51, 2.24)	(0, 0, 1.34)
ER	(0.06, 0.26, 0.67)	(0.29, 0.61, 1.35)	(0.29, 0.61, 1.35)	(0.12, 0.44, 1.08)

Step 3: *Determine the ideal and negative-ideal solutions.* Each row in the decision matrix shown in table 10.14 has a maximum rating and a minimum rating. They are the ideal and negative-ideal solutions, respectively, for the corresponding subcriterion. For arithmetic simplicity, it is assumed here that the rating with the highest most promising quantity (second parameter in the TFN) is the maximum and the rating with the lowest most promising quantity is the minimum. For example (see table 10.14), with respect to subcriterion *DD*, the maximum rating is (0.01, 0.06, 0.36) and the minimum rating is (0, 0.02, 0.18). This is because, in the row for that subcriterion, (0.01, 0.06, 0.36) is the TFN with the highest second parameter and (0, 0.02, 0.18) is the TFN with the lowest second parameter.

Step 4: *Calculate the separation distances.* The separation distances (see table 10.15) for each production facility are calculated using equations (4.52) and (4.53). For example, the positive separation distance for facility C (see table 10.15) is calculated using equation (4.52), which contains the weighted normalized fuzzy ratings of C (see table 10.14) and the ideal solution (obtained in step 3) for each subcriterion. It is important to note that because some TFNs with negative smallest possible quantities or negative most promising quantities

TABLE 10.14

Weighted normalized decision matrix (first model)

Subcriterion	Production facilities			
	A	B	C	D
DD	(0.01, 0.06, 0.36)	(0, 0.03, 0.29)	(0, 0.02, 0.18)	(0.01, 0.04, 0.36)
DC	(0, 0.02, 0.20)	(0, 0, 0.06)	(0, 0.02, 0.20)	(0, 0.02, 0.20)
DR	(0, 0.03, 0.28)	(0.01, 0.06, 0.35)	(0, 0.03, 0.28)	(0.01, 0.04, 0.35)
ES	(0, 0.06, 0.95)	(0, 0, 0.57)	(0, 0.06, 0.95)	(0, 0.06, 0.95)
CS	(0, 0, 0.18)	(0.01, 0.08, 0.59)	(0, 0.03, 0.29)	(0, 0, 0.18)
WM	(0, 0.02, 0.38)	(0, 0, 0.23)	(0, 0.04, 0.61)	(0, 0.04, 0.61)
EF	(0, 0, 0.06)	(0, 0, 0.11)	(0, 0, 0.11)	(0, 0.01, 0.11)
FU	(0, 0.01, 0.20)	(0, 0.02, 0.20)	(0, 0.01, 0.16)	(0, 0.02, 0.20)
AD	(0, 0, 0.07)	(0, 0.02, 0.23)	(0, 0.01, 0.11)	(0, 0, 0.07)
TP/SU	(0.01, 0.05, 0.20)	(0.01, 0.04, 0.16)	(0.01, 0.05, 0.20)	(0.01, 0.03, 0.11)
TP×DT	(0, 0.01, 0.08)	(0, 0.02, 0.09)	(0, 0.01, 0.05)	(0, 0.02, 0.12)
QO–QI	(0.01, 0.04, 0.19)	(0.01, 0.05, 0.22)	(0.01, 0.02, 0.10)	(0.01, 0.04, 0.16)
FC	(0.01, 0.05, 0.22)	(0.02, 0.07, 0.32)	(0.01, 0.03, 0.15)	(0.02, 0.09, 0.43)
OC	(0, 0.03, 0.18)	(0, 0.02, 0.11)	(0, 0.03, 0.17)	(0, 0.02, 0.15)
IC	(0, 0, 0.13)	(0, 0, 0.13)	(0, 0.04, 0.43)	(0, 0.02, 0.22)
IS	(0, 0, 0.06)	(0, 0.01, 0.10)	(0, 0.01, 0.12)	(0, 0.01, 0.12)
UG	(0, 0.04, 0.86)	(0, 0, 0.32)	(0, 0.03, 0.54)	(0, 0, 0.32)
ER	(0, 0.01, 0.07)	(0, 0.01, 0.15)	(0, 0.01, 0.15)	(0, 0.01, 0.12)

are obtained in this step, those TFNs are defuzzified using equation (4.8) before squaring them in the process of calculating separation distances.

Step 5: *Calculate the relative closeness to the ideal solution.* Using equation (4.54), the relative closeness coefficient for each facility in the supply chain is calculated (see table 10.16). For example, the relative closeness coefficient (i.e., 0.592) for facility C (see table 10.16) is the ratio of the facility's negative separation distance (i.e., 0.29) to the sum (i.e., 0.29 + 0.2 = 0.49) of its negative and positive separation distances (see table 10.15).

Step 6: *Rank the preference order.* Because the relative closeness coefficients of facilities A and C (0.547 and 0.592, respectively) are much higher than those of facilities B and D (0.394 and 0.434, respectively) (see table 10.16), it is evident that facilities A and C are much better than facilities B and D. If the cutoff value of the relative closeness coefficient decided by the decision maker is, say, 0.45, he or she will identify facilities A and C as the efficient ones in the region where the closed-loop supply chain is to be designed.

TABLE 10.15

Separation measures of facilities (first model)

Production facility	Positive distance S*	Negative distance S–
A	0.239	0.288
B	0.320	0.208
C	0.200	0.290
D	0.305	0.233

TABLE 10.16

Relative closeness coefficients of production facilities (first model)

Production facility	Relative closeness coefficient
A	0.547
B	0.394
C	0.592
D	0.434

10.3 Second Model (Fuzzy Logic, Extent Analysis Method, and Analytic Network Process)

In this model, the main criteria and corresponding subcriteria are the same as those in the first model (see figure 10.1); the only difference is that the *IC* and *IS* are combined and simply called *incentives*. The following numerical example illustrates the use of fuzzy logic, extent analysis method, and analytic network process in the model. Three experts carry out pair-wise comparisons among the main and subcriteria using linguistic weights (high, medium, low, etc.). These linguistic weights are converted into TFNs using table 10.17. Table 10.18 provides the comparative linguistic weights (H = high, M = medium, L = low) given to the main criteria on the second level of the hierarchy. Using fuzzy logic, these linguistic weights are converted into TFNs (using table 10.17) and then averaged to form another TFN called the average weight. For example, the average weight of criteria *ECD* with respect to *ECM* is

$$\frac{H + H + M}{3}$$

which is

TABLE 10.17

Linguistic weight conversion table for criteria and subcriteria (second model)

Linguistic weight	TFN
Very high (VH)	(0.7, 0.9, 1.0)
High (H)	(0.5, 0.7, 0.9)
Medium (M)	(0.3, 0.5, 0.7)
Low (L)	(0.1, 0.3, 0.5)
Very low (VL)	(0.0, 0.1, 0.3)

TABLE 10.18

Linguistic weights of main criteria (second model)

Criterion	ECD	ECM	AMT	POT	COS	CSE
ECD	(1, 1, 1)	(H, H, M)	(M, L, M)	(L, VL, M)	(L, VL, L)	(M, L, L)
ECM	1/(H, H, M)	(1, 1, 1)	(VH, H, H)	(H, M, H)	(M, H, M)	(H, VH, M)
AMT	1/(M, L, M)	1/(VH, H, H)	(1, 1, 1)	(H, M, H)	(M, L, M)	(M, H, L)
POT	1/(L, VL, M)	1/(H, M, H)	1/(H, M, H)	(1, 1, 1)	(H, M, H)	(M, L, H)
COS	1/(L, VL, L)	1/(M, H, M)	1/(M, L, M)	1/(H, M, H)	(1, 1, 1)	(H, M, H)
CSE	1/(M, L, L)	1/(H, VH, M)	1/(M, H, L)	1/(M, L, H)	1/(H, M, H)	(1, 1, 1)

$$\left(\frac{0.5+0.5+0.3}{3}, \frac{0.7+0.7+0.5}{3}, \frac{0.9+0.9+0.7}{3} \right) = (0.43, 0.63, 0.83)$$

(see table 10.19).

The steps of the extent analysis method are applied to the average weights to get the normalized weight vectors of main criteria. For example, consider the normalized weight vector of main criterion *ECD* shown in table 10.20; applying equations (4.9)–(4.11) to the average weights in table 10.19, the synthetic extent value of *ECD* is calculated as (0.052, 0.12, 0.181). By applying equations (4.12)–(4.14) to the synthetic extent values, the weight vectors are obtained for the main criteria, which are then normalized to get the normalized weight vectors shown in table 10.20.

Similarly, table 10.21 shows normalized weight vectors of each subcriterion with respect to its main criterion, obtained after carrying out pair-wise comparisons among subcriteria with respect to their main criteria and then applying the steps of the extent analysis method.

Table 10.22 shows the matrix of interdependencies obtained after carrying out pair-wise comparisons among the subcriteria and carrying out the steps involved in the extent analysis method.

The super matrix M is made to converge to obtain a long-term stable set of weights. For convergence, M must be made column stochastic, which is done

TABLE 10.19

Average weights of main criteria (second model)

Criterion	ECD	ECM	AMT	POT	COS	CSE
ECD	(1, 1, 1)	(0.43, 0.63, 0.83)	(0.23, 0.43, 0.63)	(0.14, 0.3, 0.5)	(0.07, 0.23, 0.43)	(0.17, 0.37, 0.57)
ECM	(1.20, 1.58, 2.32)	(1, 1, 1)	(0.57, 0.77, 0.63)	(0.43, 0.63, 0.83)	(0.37, 0.57, 0.77)	(0.5, 0.7, 0.57)
AMT	(1.58, 2.32, 4.34)	(1.58, 1.29, 1.75)	(1, 1, 1)	(0.43, 0.63, 0.83)	(0.23, 0.43, 0.63)	(0.3, 0.5, 0.7)
POT	(2, 3.33, 7.14)	(1.2, 1.58, 2.32)	(1.2, 1.58, 2.32)	(1, 1, 1)	(0.43, 0.63, 0.83)	(0.3, 0.5, 0.7)
COS	(2.32, 4.34, 4.28)	(1.29, 0.07, 2.7)	(1.58, 2.32, 4.34)	(1.2, 1.58, 2.32)	(1, 1, 1)	(0.43, 0.63, 0.83)
CSE	(1.75, 2.7, 5.88)	(1.75, 1.42, 2)	(1.42, 2, 3.33)	(1.42, 2, 3.33)	(1.2, 1.58, 2.32)	(1, 1, 1)
Sum	(9.87, 15.29, 34.9)	(7.27, 6.01, 10.6)	(6.02, 8.11, 12.2)	(4.63, 6.14, 8.81)	(3.3, 4.44, 5.98)	(2.7, 3.7, 4.37)

TABLE 10.20

Weights of main criteria (P_j) (second model)

Criterion	Weight
ECD	0.008
ECM	0.093
AMT	0.159
POT	0.223
COS	0.253
CSE	0.260

by raising M to the power of 2^{k+1}, where k is an arbitrarily large number; in this example, $k = 59$. Table 10.23 shows the converged super matrix (A_{kj}^I).

Table 10.24 shows the linguistic ratings and their corresponding TFNs, which are used to evaluate the production facilities with respect to each subcriterion.

Relative weights of the production facilities are obtained by carrying out pair-wise comparisons among the production facilities with respect to the subcriteria and applying the steps involved in the extent analysis method. Table 10.25 shows the relative weights of the production facilities (S_{ikj}).

Using equation (4.2) of analytic network process (ANP), the desirability indices (*DI*) for each production facility are calculated and shown in table 10.26.

The overall performance index for each production facility is calculated by multiplying the desirability index (table 10.26) of each production facility for each criterion by the weight of the criterion (table 10.20), summing over

TABLE 10.21

Weights of subcriteria with respect to main criteria (A_{kj}^{D}) (second model)

Subcriterion	Weight
DD	0.365
DC	0.333
DR	0.302
ES	0.057
CS	0.290
WM	0.653
EF	0.067
FU	0.335
AD	0.598
TP/SU	0.136
TP×DT	0.334
QO–QI	0.527
FC	0.121
OC	0.879
Incentives	0.097
UG	0.477
ER	0.427

all the criteria, and normalizing those weighted sums. Table 10.27 shows the overall performance indices for the four production facilities.

Facility S's overall performance index is the largest; hence, it is the best of the lot.

10.4 Third Model (Fuzzy Multicriteria Analysis Method)

In this model, the main criteria and corresponding subcriteria are the same as those in the second model. The model is illustrated using a numerical example, as follows. Table 10.28 shows the TFNs used for making qualitative assessments (pair-wise comparisons). Table 10.29 shows the pair-wise comparisons among the main criteria with respect to the goal of evaluation of production facilities using the membership functions shown in table 10.28. By applying the extent analysis method on the matrix of pair-wise comparisons (table 10.29), the corresponding weights of the main criteria are calculated and shown in table 10.30.

Similarly, pair-wise comparisons are carried out among the candidate production facilities with respect to the subcriteria, and the extent analysis method is applied on each of those matrices to derive the corresponding

TABLE 10.22

Matrix of interdependencies (super matrix M) (second model)

	DD	DC	DR	ES	CS	WM	EF	FU	AD	TP/SU	TP×DT	QO–QI	FC	OC	Incentives	UG	ER
DD	0	0.2	0.07	0	0	0	0	0	0	0	0	0	0	0	0	0	0
DC	0.11	0	0.92	0	0	0	0	0	0	0	0	0	0	0	0	0	0
DR	0.88	0.79	0	0	0	0	0	0	0	0	0	0	0	0	0	0	0
ES	0	0	0	0	0.11	0.14	0	0	0	0	0	0	0	0	0	0	0
CS	0	0	0	0.51	0	0.85	0	0	0	0	0	0	0	0	0	0	0
WM	0	0	0	0.48	0.88	0	0	0	0	0	0	0	0	0	0	0	0
EF	0	0	0	0	0	0	0	0.5	0.33	0	0	0	0	0	0	0	0
FU	0	0	0	0	0	0	0.51	0	0.66	0	0	0	0	0	0	0	0
AD	0	0	0	0	0	0	0.48	0.46	0	0	0	0	0	0	0	0	0
TP/SU	0	0	0	0	0	0	0	0	0	0	0.5	0.25	0	0	0	0	0
TP×DT	0	0	0	0	0	0	0	0	0	0.11	00.4	0.74	0	0	0	0	0
QO–QI	0	0	0	0	0	0	0	0	0	0.88	0	0	0	0	0	0	0
FC	0	0	0	0	0	0	0	0	0	0	0	0	0	0.5	0	0	0
OC	0	0	0	0	0	0	0	0	0	0	0	0	0.5	0	0	0	0
Incentives	0	0	0	0	0	0	0	0	0	0	0	0	0	0	0	0.2	0.33
UG	0	0	0	0	0	0	0	0	0	0	0	0	0	0	0.11	0	0.66
ER	0	0	0	0	0	0	0	0	0	0	0	0	0	0	0.88	0.79	0

TABLE 10.23

Converged super matrix (second model)

Subcriterion	Weight
DD	0.121
DC	0.427
DR	0.45
ES	0.11
CS	0.44
WM	0.44
EF	0.29
FU	0.37
AD	0.32
TP/SU	0.27
TP×DT	0.32
QO–QI	0.4
FC	7.89E-31
OC	7.89E-31
Incentives	0.21
UG	0.32
ER	0.45

TABLE 10.24

Linguistic weight conversion table for production facilities (second model)

Linguistic weight	TFN
Very good (VG)	(7, 10, 10)
Good (G)	(5, 7, 10)
Fair (F)	(2, 5, 8)
Poor (P)	(1, 3, 5)
Very poor (VP)	(1, 1, 3)

fuzzy weights. In this example, there are seventeen subcriteria that result in seventeen such pair-wise comparisons among the candidate production facilities. For example, tables 10.31 and 10.32 show the pair-wise comparisons among the candidate production facilities with respect to the subcriteria design for disassembly (*DD*) and design for recycling (*DC*), respectively.

Table 10.33 shows the decision matrix (*X*), the matrix showing the performance of each candidate production facility with respect to the subcriteria. These weights are obtained by applying the extent analysis method on each of the pair-wise comparison matrices of candidate production facilities with respect to the subcriteria.

Pair-wise comparisons are carried out among the subcriteria with respect to their main criteria. The fuzzy weights of the subcriteria with respect to the

TABLE 10.25

Relative weights of production facilities with respect to subcriteria (second model)

Subcriteria/ production facilities	P	Q	R	S
DD	0.26	0.26	0.24	0.24
DC	0.26	0.25	0.24	0.24
DR	0.26	0.26	0.24	0.24
ES	0.26	0.26	0.24	0.24
CS	0.26	0.25	0.25	0.24
WM	0.26	0.26	0.24	0.24
EF	0.26	0.25	0.25	0.24
FU	0.26	0.26	0.23	0.25
AD	0.26	0.25	0.25	0.24
TP/SU	0.26	0.25	0.25	0.24
TP×DT	0.25	0.27	0.24	0.24
QO–QI	0.26	0.25	0.25	0.24
FC	0.26	0.23	0.26	0.25
OC	0.26	0.23	0.26	0.25
Incentives	0.26	0.25	0.25	0.24
UG	0.26	0.25	0.24	0.25
ER	0.26	0.25	0.25	0.24

TABLE 10.26

Desirability indices (*DI*) (second model)

Criterion	P	Q	R	S
ECD	0.09	0.08	0.08	0.08
ECM	0.11	0.11	0.10	0.10
AMT	0.09	0.09	0.08	0.08
POT	0.09	0.09	0.09	0.15
COS	0.00	0.00	0.00	0.00
CSE	0.10	0.09	0.09	0.09

TABLE 10.27

Overall performance indices for production facilities (second model)

Production facility	Overall performance index
P	0.26
Q	0.20
R	0.19
S	0.35

TABLE 10.28

Triangular fuzzy numbers used for qualitative assessments (third model)

Fuzzy number	Membership function
1	(1, 1, 3)
x	(x − 2, x, x + 2) for x = 3, 5, 7
9	(7, 9, 11)

TABLE 10.29

Pair-wise comparisons among main criteria (third model)

	ECD	ECM	AMT	POT	COS	CSE
ECD	(1, 1, 3)	(1, 3, 5)	(7, 9, 11)	(5, 7, 9)	(1, 3, 5)	(1, 3, 5)
ECM	1/(1, 3, 5)	(1, 1, 3)	(1, 3, 5)	(7, 9, 11)	(1, 1, 3)	(5, 7, 9)
AMT	1/(7, 9, 11)	1/(1, 3, 5)	(1, 1, 3)	(3, 5, 7)	(5, 7, 9)	(7, 9, 11)
POT	1/(5, 7, 9)	1/(7, 9, 11)	1/(3, 5, 7)	(1, 1, 3)	(7, 9, 11)	(1, 3, 5)
COS	1/(1, 3, 5)	1/(1, 1, 3)	1/(5, 7, 9)	1/(7, 9, 11)	(1, 1, 3)	(5, 7, 9)
CSE	1/(1, 3, 5)	1/(5, 7, 9)	1/(7, 9, 11)	1/(1, 3, 5)	1/(5, 7, 9)	(1, 1, 3)

TABLE 10.30

Fuzzy weights of main criteria (third model)

Criterion	TFN
ECD	(0.12, 0.29, 0.62)
ECM	(0.11, 0.23, 0.52)
AMT	(0.12, 0.25, 0.51)
POT	(0.05, 0.13, 0.29)
COS	(0.03, 0.08, 0.20)
CSE	(0.01, 0.02, 0.09)

TABLE 10.31

Pair-wise comparisons among production facilities with respect to *DD* (third model)

	P1	P1	P3	P4
P1	(1, 1, 3)	(3, 5, 7)	(1, 3, 5)	(5, 7, 9)
P2	1/(3, 5, 7)	(1, 1, 3)	(5, 7, 9)	(3, 5, 7)
P3	1/(1, 3, 5)	1/(5, 7, 9)	(1, 1, 3)	(1, 3, 5)
P4	1/(5, 7, 9)	1/(3, 5, 7)	1/(1, 3, 5)	(1, 1, 3)

TABLE 10.32

Pair-wise comparisons among production facilities with respect to *DC* (third model)

	P1	P1	P3	P4
P1	(1, 1, 3)	(1, 3, 5)	(5, 7, 9)	(7, 9, 11)
P2	1/(1, 3, 5)	(1, 1, 3)	(3, 5, 7)	(5, 7, 9)
P3	1/(5, 7, 9)	1/(3, 5, 7)	(1, 1, 3)	(1, 3, 5)
P4	1/(7, 9, 11)	1/(5, 7, 9)	1/(1, 3, 5)	(1, 1, 3)

TABLE 10.33

Fuzzy decision matrix (*X*) (third model)

	P1	P2	P3	P4
DD	(0.17, 0.45, 1.04)	(0.16, 0.37, 0.84)	(0.04, 0.12, 0.4)	(0.025, 0.04, 0.19)
DC	(0.22, 0.5, 1.04)	(0.15, 0.33, 0.74)	(0.03, 0.11, 0.31)	(0.02, 0.04, 0.16)
DR	(0.12, 0.34, 0.76)	(0.17, 0.36, 0.74)	(0.13, 0.25, 0.52)	(0.02, 0.03, 0.12)
ES	(0.17, 0.42, 0.88)	(0.16, 0.35, 0.77)	(0.07, 0.16, 0.41)	(0.02, 0.04, 0.17)
CS	(0.14, 0.39, 0.96)	(0.12, 0.31, 0.75)	(0.1, 0.23, 0.54)	(0.02, 0.05, 0.22)
WM	(0.38, 0.82, 1.64)	(0.35, 0.78, 1.45)	(0.2, 0.47, 0.91)	(0.04, 0.07, 0.25)
EF	(0.38, 0.96, 2.08)	(0.22, 0.60, 1.44)	(0.13, 0.33, 0.84)	(0.04, 0.08, 0.35)
FU	(0.19, 0.45, 0.96)	(0.15, 0.33, 0.72)	(0.06, 0.15, 0.38)	(0.02, 0.04, 0.17)
AD	(0.18, 0.41, 0.84)	(0.14, 0.3, 0.65)	(0.12, 0.24, 0.47)	(0.02, 0.03, 0.11)
TP/SU	(0.38, 0.82, 1.64)	(0.35, 0.78, 1.45)	(0.2, 0.47, 0.91)	(0.04, 0.07, 0.25)
TP×DT	(0.17, 0.45, 1.04)	(0.16, 0.37, 0.84)	(0.04, 0.12, 0.4)	(0.02, 0.04, 0.19)
QO–QI	(0.17, 0.45, 1.04)	(0.16, 0.37, 0.84)	(0.04, 0.12, 0.4)	(0.02, 0.04, 0.19)
FC	(0.12, 0.32, 0.71)	(0.2, 0.39, 0.75)	(0.12, 0.24, 0.49)	(0.02, 0.03, 0.11)
OC	(0.22, 0.46, 0.91)	(0.14, 0.3, 0.62)	(0.09, 0.19, 0.4)	(0.022, 0.03, 0.12)
Incentives	(0.75, 2.22, 4.66)	(0.53, 1.55, 3.37)	(0.62, 1.45, 2.95)	(0.1, 0.22, 0.84)
UG	(0.22, 0.46, 0.91)	(0.14, 0.3, 0.62)	(0.09, 0.19, 0.4)	(0.022, 0.03, 0.12)
ER	(0.17, 0.45, 1.04)	(0.16, 0.37, 0.84)	(0.04, 0.12, 0.4)	(0.02, 0.04, 0.19)

main criteria are obtained by applying the extent analysis method to those pair-wise comparison matrices. Table 10.34 shows the fuzzy weights of the subcriteria with respect to their main criteria.

Table 10.35 shows the fuzzy reciprocal judgment matrix (*W*) that is obtained by multiplying the fuzzy weights of the subcriteria by the corresponding main criteria weights.

A fuzzy performance matrix *Z* representing the overall performance of all candidate production facilities with respect to each criterion is obtained by multiplying the weight vector (fuzzy reciprocal judgment matrix, *W*) by the decision matrix, *X*. Table 10.36 shows the fuzzy performance matrix, *Z*.

An interval performance matrix is derived by using an α-cut on the performance matrix, *Z*, where $0 \leq \alpha \leq 1$. The value of α represents the decision

TABLE 10.34

Fuzzy weights of subcriteria with respect to main criteria
(third model)

Subcriterion	Fuzzy weight
DD	(0.14, 0.43, 1.03)
DC	(0.24, 0.49, 0.99)
DR	(0.03, 0.06, 0.28)
ES	(0.16, 0.53, 1.43)
CS	(0.13, 0.37, 0.98)
WM	(0.04, 0.09, 0.41)
EF	(0.28, 0.56, 1.16)
FU	(0.12, 0.31, 0.69)
AD	(0.04, 0.11, 0.29)
TP/SU	(0.15, 0.48, 1.2)
TP×DT	(0.19, 0.44, 1.04)
QO–QI	(0.039, 0.07, 0.28)
FC	(0.3, 0.83, 1.94)
OC	(0.085, 0.16, 0.64)
Incentives	(0.15, 0.48, 1.2)
UG	(0.19, 0.44, 1.04)
ER	(0.03, 0.07, 0.28)

TABLE 10.35

Fuzzy reciprocal judgment matrix (*W*) (third model)

Subcriterion	Fuzzy weight
DD	(0.017, 0.12, 0.64)
DC	(0.02, 0.14, 0.615)
DR	(0.004, 0.019, 0.177)
ES	(0.01, 0.12, 0.74)
CS	(0.015, 0.08, 0.514)
WM	(0.005, 0.02, 0.215)
EF	(0.03, 0.14, 0.59)
FU	(0.015, 0.07, 0.35)
AD	(0.005, 0.028, 0.151)
TP/SU	(0.008, 0.06, 0.34)
TP×DT	(0.01, 0.05, 0.3)
QO–QI	(0.002, 0.009, 0.08)
FC	(0.01, 0.06, 0.39)
OC	(0.002, 0.01, 0.13)
Incentives	(0.002, 0.011, 0.028)
UG	(0.002, 0.01, 0.024)
ER	(0.0005, 0.0001, 0.0006)

TABLE 10.36

Fuzzy performance matrix (Z) (third model)

	P1	P1	P3	P4
DD	**(0.003, 0.056, 0.674)**	(0.002, 0.04, 0.54)	(0.0007, 0.015, 0.25)	(0.0004, 0.005, 0.12)
DC	(0.006, 0.07, 0.64)	(0.004, 0.04, 0.45)	(0.001, 0.01, 0.195)	(0.0006, 0.005, 0.09)
DR	(0.0005, 0.006, 0.13)	(0.0008, 0.007, 0.13)	(0.0006, 0.005, 0.09)	(0.00009, 0.0007, 0.022)
ES	(0.003, 0.05, 0.65)	(0.003, 0.04, 0.57)	(0.001, 0.02, 0.31)	(0.0004, 0.005, 0.12)
CS	(0.002, 0.03, 0.49)	(0.001, 0.02, 0.38)	(0.001, 0.02, 0.281)	(0.004, 0.004, 0.116)
WM	(0.001, 0.017, 0.35)	(0.001, 0.016, 0.312)	(0.001, 0.01, 0.19)	(0.0002, 0.001, 0.05)
EF	(0.012, 0.135, 1.23)	(0.007, 0.08, 0.85)	(0.004, 0.04, 0.5)	(0.001, 0.01, 0.212)
FU	(0.003, 0.03, 0.34)	(0.002, 0.02, 0.25)	(0.001, 0.01, 0.136)	(0.0003, 0.003, 0.06)
AD	(0.001, 0.01, 0.12)	(0.0008, 0.008, 0.09)	(0.0007, 0.006, 0.07)	(0.0001, 0.0009, 0.017)
TP/SU	(0.003, 0.03, 0.57)	(0.002, 0.04, 0.5)	(0.001, 0.02, 0.31)	(0.0003, 0.004, 0.08)
TP×DT	(0.001, 0.02, 0.31)	(0.001, 0.02, 0.25)	(0.0004, 0.007, 0.12)	(0.0002, 0.002, 0.35)
QO–QI	(0.0003, 0.0004, 0.08)	(0.0003, 0.003, 0.06)	(0.00008, 0.001, 0.032)	(0.00005, 0.0004, 0.016)
FC	(0.001, 0.022, 0.28)	(0.002, 0.02, 0.29)	(0.001, 0.01, 0.19)	(0.0002, 0.002, 0.04)
OC	(0.0006, 0.006, 0.11)	(0.0004, 0.004, 0.08)	(0.0002, 0.002, 0.05)	(0.00006, 0.0005, 0.01)
Incentives	(0.001, 0.02, 0.13)	(0.001, 0.017, 0.09)	(0.001, 0.016, 0.082)	(0.0002, 0.002, 0.02)
UG	(0.0005, 0.004, 0.022)	(0.0003, 0.003, 0.015)	(0.0002, 0.002, 0.009)	(0.00005, 0.0003, 0.003)
ER	(0.00008, 0.0007, 0.006)	((0.00008, 0.0006, 0.005)	(0.00002, 0.0002, 0.002)	(0.00001, 0.00007, 0.001)

maker's degree of confidence in his or her fuzzy assessments regarding the production facility ratings and criteria weights. The larger the value of α, the more confident the decision maker is about the fuzzy assessments, i.e., the assessments are closer to the most possible value a_2 of the triangular fuzzy number (a_1, a_2, a_3). In this example, $\alpha = 0.5$. For example, consider the fuzzy performance rating of candidate production facility P1 (shown in bold in table 10.36) with respect to subcriterion *DD*. The α-cut on this performance rating can be performed as $((0.003)^{0.5}, (0.674)^{0.5}) = (0.055, 0.82)$. Table 10.37 shows the interval performance matrix, Z_α, obtained by performing an α-cut on the matrix shown in table 10.36.

An overall crisp performance matrix that incorporates the decision maker's attitude toward risk, using an optimism index λ ($\lambda = 1$ implies the decision maker has an optimistic view, 0 implies a pessimistic view, and 0.5 implies a moderate view), is calculated using equations (4.22) and (4.23). Table 10.38 shows the crisp performance matrix at an optimism index $\lambda = 0.5$. For example, consider the α-cut on the performance rating (shown in bold in table 10.37) of

TABLE 10.37

Interval performance matrix, Z_α (third model)

	P1	P2	P3	P4
DD	**(0.055, 0.82)**	(0.05, 0.73)	(0.02, 0.5)	(0.02, 0.35)
DC	(0.08, 0.8)	(0.06, 0.67)	(0.032, 0.44)	(0.025, 0.31)
DR	(0.024, 0.368)	(0.028, 0.36)	(0.024, 0.3)	(0.009, 0.15)
ES	(0.05, 0.81)	(0.05, 0.76)	(0.03, 0.55)	(0.02, 0.36)
CS	(0.04, 0.7)	(0.044, 0.62)	(0.04, 0.53)	(0.02, 0.34)
WM	(0.043, 0.596)	(0.04, 0.55)	(0.03, 0.44)	(0.014, 0.233)
EF	(0.11, 1.11)	(0.08, 0.92)	(0.06, 0.7)	(0.03, 0.46)
FU	(0.055, 0.58)	(0.04, 0.504)	(0.032, 0.369)	(0.019, 0.247)
AD	(0.032, 0.357)	(0.028, 0.313)	(0.026, 0.267)	(0.01, 0.131)
TP/SU	(0.05, 0.75)	(0.05, 0.7)	(0.04, 0.56)	(0.02, 0.29)
TP×DT	(0.042, 0.561)	(0.041, 0.503)	(0.02, 0.347)	(0.016, 0.243)
QO–QI	(0.019, 0.292)	(0.018, 0.262)	(0.009, 0.181)	(0.007, 0.127)
FC	(0.03, 0.53)	(0.04, 0.54)	(0.03, 0.43)	(0.01, 0.21)
OC	(0.02, 0.345)	(0.02, 0.285)	(0.017, 0.23)	(0.008, 0.128)
Incentives	(0.03, 0.36)	(0.03, 0.3)	(0.03, 0.28)	(0.014, 0.15)
UG	(0.023, 0.148)	(0.018, 0.123)	(0.015, 0.099)	(0.007, 0.055)
ER	(0.009, 0.083)	(0.009, 0.07)	(0.004, 0.051)	(0.003, 0.036)

TABLE 10.38

Crisp performance matrix (third model)

	P1	P2	P3	P4
DD	**0.57**	0.54	0.44	0.37
DC	0.36	0.42	0.54	0.59
DR	0.38	0.39	0.36	0.24
ES	0.57	0.55	0.47	0.37
CS	0.53	0.50	0.47	0.36
WM	0.49	0.48	0.42	0.30
EF	0.70	0.63	0.55	0.44
FU	0.50	0.47	0.39	0.32
AD	0.39	0.36	0.34	0.23
TP/SU	0.55	0.54	0.48	0.34
TP×DT	0.48	0.46	0.37	0.31
QO–QI	0.34	0.32	0.26	0.22
FC	0.46	0.48	0.43	0.29
OC	0.37	0.34	0.31	0.22
Incentives	0.40	0.37	0.36	0.26
UG	0.27	0.24	0.22	0.16
ER	0.19	0.18	0.15	0.13

production facility P1 with respect to subcriterion DD. Using equation (4.23), the crisp performance rating is $(0.5\times0.82^{0.5}+0.5\times0.055^{0.5})=0.57$.

The crisp performance matrix is normalized using equation (4.25) to obtain the normalized performance matrix. Table 10.39 shows the normalized performance matrix. For example, the normalized performance score of production facility P1 (shown in bold in table 10.39) with respect to subcriterion DD is calculated using equation (4.25) as

$$\frac{(0.5707)}{\sqrt{0.5707^2+0.5442^2+0.438^2+0.3713^2}}=0.585$$

(Note that the values 0.5707, 0.5442, 0.438, and 0.3713 are in the first row of table 10.38.)

Using equations (4.26) and (4.27), the positive- and negative-ideal solutions are calculated by selecting maximum and minimum ratings across all candidate production facilities. Table 10.40 shows the positive- and negative-ideal solutions. For example, the positive- and negative-ideal solutions across all candidate production facilities with respect to DC are the maximum and minimum values, respectively, from the second row of table 10.39.

By applying the vector-matching function, the degree of similarity (see table 10.41) between the normalized performance score of each candidate

TABLE 10.39
Normalized performance matrix (third model)

	P1	P2	P3	P4
DD	**0.585**	0.558	0.449	0.381
DC	0.606	0.554	0.434	0.371
DR	0.550	0.557	0.513	0.352
ES	0.540	0.524	0.445	0.354
CS	0.536	0.509	0.474	0.370
WM	0.544	0.528	0.469	0.335
EF	0.593	0.536	0.469	0.374
FU	0.588	0.547	0.464	0.375
AD	0.577	0.540	0.505	0.346
TP/SU	0.571	0.555	0.492	0.352
TP×DT	0.585	0.558	0.449	0.380
QO–QI	0.585	0.558	0.449	0.381
FC	0.546	0.567	0.508	0.349
OC	0.593	0.537	0.484	0.355
Incentives	0.569	0.525	0.517	0.366
UG	0.593	0.537	0.484	0.354
ER	0.586	0.559	0.447	0.380

TABLE 10.40

Positive- and negative-ideal solutions (third model)

Subcriterion	Positive-ideal solution	Negative-ideal solution
DD	0.585	0.381
DC	**0.606**	**0.371**
DR	0.557	0.352
ES	0.540	0.354
CS	0.536	0.370
WM	0.544	0.335
EF	0.593	0.374
FU	0.588	0.375
AD	0.577	0.346
TP/SU	0.571	0.352
TP×DT	0.585	0.380
QO–QI	0.585	0.381
FC	0.567	0.349
OC	0.593	0.355
Incentives	0.569	0.366
UG	0.593	0.354
ER	0.586	0.380

TABLE 10.41

Degree of similarity to positive- and negative-ideal solutions (third model)

Production facility	Degree of similarity to positive-ideal solution	Degree of similarity to negative-ideal solution
P1	0.997	1.576
P2	0.946	1.495
P3	0.822	1.290
P4	0.646	1.013

production facility and the positive- and negative-ideal solutions can be calculated using equation (4.28).

A preferred candidate production facility should have a higher degree of similarity to the positive-ideal solution and a lower degree of similarity to the negative-ideal solution. Hence, an overall performance index for each candidate facility with the decision maker's α level of confidence and λ degree of optimism toward risk is determined using equation (4.29). Table 10.42 shows the overall performance indices of the four candidate production facilities. The larger the performance index, the better the candidate. For example, for facility P3, the overall performance index is calculated using equation (4.29) as

$$\frac{(0.822)}{(0.822)+(1.290)}=0.389$$

TABLE 10.42
Overall performance indices (third model)

Production facility	Overall performance index
P1	0.388
P2	0.387
P3	**0.389**
P4	0.390

Because production facility P4's overall performance index is the largest, it is the best of the lot.

10.5   Conclusions

In this chapter, three models to evaluate production facilities operating in a region where a closed-loop supply chain is to be designed were presented. These models evaluate production facilities in terms of both environmental consciousness and potentiality. The first model employs fuzzy logic and technique for order preference by similarity to ideal solution (TOPSIS); the second model employs fuzzy logic, extent analysis method, and analytic network process; the third model employs fuzzy multicriteria analysis method.

References

1. Gungor, A., and Gupta, S. M. 1999. Issues in environmentally conscious manufacturing and product recovery: A survey. *Computers and Industrial Engineering* 36:811–53.

11

Evaluation of Futurity of Used Products

11.1 The Issue

A major driver for companies interested in collecting used products is recoverable value through reprocessing (remanufacturing or recycling). However, the companies seldom know when those products were bought and why they were discarded. Also, the products do not indicate their remaining life periods. Hence, they often undergo partial or complete disassembly for subsequent reprocessing. The authors are of the opinion that for some used products, it might make more sense to make necessary repairs to the products and sell them on secondhand markets than to disassemble them for subsequent reprocessing. To this end, in this chapter, using a numerical example, it is shown how an expert system can be built using Bayesian updating and fuzzy logic to decide whether it is sensible to repair a used product of interest for subsequent sale on a secondhand market. It is assumed here that the used product of interest functions improperly; it is obviously sensible to sell a properly functioning used product on a secondhand market.

Consider the used product shown in figure 11.1. Given that this product is not functioning properly, one shall decide whether it is sensible to repair it for subsequent sale on a secondhand market. Table 11.1 shows the probability values used to implement Bayesian updating.

Section 11.2 presents the procedure to use fuzzy logic in Bayesian updating. Section 11.3 gives the list of the rules used in Bayesian updating. Section 11.4 presents the implementation of Bayesian updating. Section 11.5 presents the employment of FLEX shell [1] to build an expert system that can be used to decide whether it is sensible to repair an improperly functioning used product of interest for subsequent sale on a secondhand market. Finally, section 11.6 gives some conclusions.

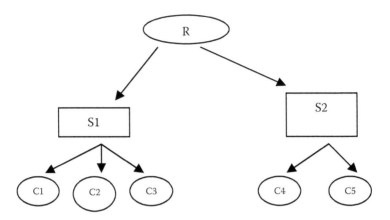

FIGURE 11.1
Used product.

TABLE 11.1

Probability values used in Bayesian updating

H	E	P(H)	O(H)	P(E∣H)	P(E∣~H)	A	D
S1 needs repair	Product needs repair	0.60	1.50	1.00	0.60	1.67	0.00
S2 needs repair	Product needs repair	0.70	2.33	1.00	0.40	2.50	0.00
C1 needs repair	S1 needs repair	0.45	0.82	1.00	0.45	2.22	0.00
C2 needs repair	S1 needs repair	0.55	1.22	1.00	0.30	3.33	0.00
C3 needs repair	S1 needs repair	0.30	0.43	1.00	0.55	1.82	0.00
C4 needs repair	S2 needs repair	0.32	0.47	1.00	0.70	1.43	0.00
C5 needs repair	S2 needs repair	0.10	0.11	1.00	0.80	1.25	0.00
Sensible to repair product	C1 needs repair	0.60	1.50	**0.70**	**0.20**	3.50	0.38
Sensible to repair product	C2 needs repair	0.60	1.50	**0.60**	**0.30**	2.00	0.57
Sensible to repair product	C3 needs repair	0.60	1.50	**0.45**	**0.60**	0.75	1.38
Sensible to repair product	C4 needs repair	0.60	1.50	**0.10**	**0.75**	0.13	3.60
Sensible to repair product	C5 needs repair	0.60	1.50	**0.85**	**0.40**	2.13	0.25

11.2   Usage of Fuzzy Logic

Because it is difficult for an expert to guess the probabilities shown in bold (unlike the rest) in table 11.1, fuzzy logic is used to calculate them as follows:

1. Ask the expert to assign a linguistic rating to $P(E|H)$ for each component in the used product with respect to each of the following factors (see table 11.2):

   a. Is it economical to repair or replace the component (more economical implies higher rating)?

   b. If disposed of, will the component be harmful to the environment (more harmful implies higher rating)?

   c. What is the remaining life period of the component (longer life implies higher rating)?

   d. Is the raw material used to make the component depleting quickly (faster depletion implies higher rating)?

   e. Is it difficult to repair the component (more difficult implies lower rating)?

2. Use the data in table 11.3 to convert the linguistic ratings into TFNs.

3. Calculate the average fuzzy $P(E|H)$ value for each component.

TABLE 11.2

Linguistic $P(E|H)$ ratings

	a	b	c	d	e
C1	High	High	Medium	Medium	High
C2	Very high	High	Very high	Medium	Low
C3	Low	Low	Very low	Very high	Medium
C4	High	High	High	High	Very low
C5	Medium	High	High	High	Medium

TABLE 11.3

Conversion table for linguistic $P(E|H)$ ratings in Bayesian updating

Linguistic rating	Triangular fuzzy number
Very high (VH)	(0.7, 0.9, 1.0)
High (H)	(0.5, 0.7, 0.9)
Medium (M)	(0.3, 0.5, 0.7)
Low (L)	(0.1, 0.3, 0.5)
Very low (VL)	(0.0, 0.1, 0.3)

4. Defuzzify the average $P(E|H)$ for each component using equation (4.8).

Apply steps 1–4 to calculate appropriate $P(E|\sim H)$ values for each component. In order to be saved from tedious calculations, it is assumed here that the values shown in bold in table 11.1 are the defuzzified average probabilities obtained *after* performing steps 1–4. For the sake of clarity, however, the calculation procedure is explained using an example below.

Suppose that one wishes to calculate the $P(E|H)$ value (numerical) for a component in a used product. The four steps are implemented as follows:

1. The expert linguistically rates the component with respect to factors a–e as very high, high, medium, medium, and low, respectively.
2. Using table 11.3, the linguistic ratings are converted into TFNs.
3. The average fuzzy $P(E|H)$ is equal to

$$\left(\frac{0.7+0.5+0.3+0.3+0.1}{5}, \frac{0.9+0.7+0.5+0.5+0.3}{5}, \frac{1.0+0.9+0.7+0.7+0.5}{5} \right),$$

i.e., (0.38, 0.58, 0.76).

4. Defuzzifying the average fuzzy $P(E|H)$ using equation 4.8, one gets

$$\frac{(0.76-0.38)+(0.58-0.38)}{3}+0.38 = 0.57$$

11.3   Rules Used in Bayesian Updating

Rule 1: IF product needs repair (AFFIRMS: 1.67; DENIES: 0.00), THEN S1 needs repair.

Rule 2: IF product needs repair (AFFIRMS: 2.50; DENIES: 0.00), THEN S2 needs repair.

Rule 3: IF S1 needs repair (AFFIRMS: 2.22; DENIES: 0.00), THEN C1 needs repair.

Rule 4: IF S1 needs repair (AFFIRMS: 3.33; DENIES: 0.00), THEN C2 needs repair.

Rule 5: IF S1 needs repair (AFFIRMS: 1.82; DENIES: 0.00), THEN C3 needs repair.

Rule 6: IF S2 needs repair (AFFIRMS: 1.43; DENIES: 0.00), THEN C4 needs repair.

Rule 7: IF S2 needs repair (AFFIRMS: 1.25; DENIES: 0.00), THEN C5 needs repair.

Rule 8: IF C1 needs repair (AFFIRMS: 3.50; DENIES 0.38) AND C2 needs repair (AFFIRMS: 2.00; DENIES 0.57) AND C3 needs repair (AFFIRMS: 0.75; DENIES 1.38) AND C4 needs repair (AFFIRMS: 0.13; DENIES 3.60) AND C5 needs repair (AFFIRMS: 2.13; DENIES 0.25), THEN it is sensible to repair the product.

11.4   Bayesian Updating

Refer to table 11.1 while reading this section.

Rule 1: H = S1 needs repair; O(H) = 1.50; E = product needs repair; A = 1.67; O(H|E) = O(H)×(A) = 2.51.

Rule 2: H = S2 needs repair; O(H) = 2.33; E = product needs repair; A = 2.50; O(H|E) = O(H)×(A) = 5.83.

Rule 3: H = C1 needs repair; O(H) = 0.82; E = S1 needs repair; O(E) = 2.51; P(E) = 0.72; A = 2.22; A′ = [2(A − 1)×P(E)] + 2 − A = 1.54; O(H|E) = O(H)×(A′) = (0.82)×(1.54) = 1.26.

Rule 4: H = C2 needs repair; O(H) = 1.22; E = S1 needs repair; O(E) = 2.51; P(E) = 0.72; A = 3.33; A′ = [2(A − 1)×P(E)] + 2 − A = 2.03; O(H|E) = O(H)×(A′) = (1.22)×(2.03) = 2.48.

Rule 5: H = C3 needs repair; O(H) = 0.43; E = S1 needs repair; O(E) = 2.51; P(E) = 0.72; A = 1.82; A′ = [2(A − 1)×P(E)] + 2 − A = 1.36; O(H|E) = O(H)×(A′) = (0.43)×(1.36) = 0.58.

Rule 6: H = C4 needs repair; O(H) = 0.47; E = S2 needs repair; O(E) = 5.83; P(E) = 0.85; A = 1.43; A′ = [2(A − 1)×P(E)] + 2 − A = 1.30; O(H|E) = O(H)×(A′) = (0.47)×(1.30) = 0.61.

Rule 7: H = C5 needs repair; O(H) = 0.11; E = S2 needs repair; O(E) = 5.83; P(E) = 0.85; A = 1.25; A′ = [2(A − 1)×P(E)] + 2 − A = 1.18; O(H|E) = O(H)×(A′) = (0.11)×(1.18) = 0.13.

Rule 8: H = sensible to repair product; O(H) = 1.50;

E1 = C1 needs repair; O(E1) = 1.26; P(E1) = 0.56; A1 = 3.50; A1′ = [2·(A1 − 1)×P(E1)] + 2 − A1 = 1.30;

E2 = C2 needs repair; O(E2) = 2.48; P(E2) = 0.71; A2 = 2.00; A2′ = [2·(A2 − 1)×P(E2)] + 2 − A2 = 1.42;

E3 = C3 needs repair; O(E3) = 0.58; P(E3) = 0.37; D3 = 1.38; D3′ = [2·(1 − D3)×P(E3)] + D3 = 1.09;

E4 = C4 needs repair; O(E4) = 0.61; P(E4) = 0.38; D4 = 3.60; D4′ = [2·(1 − D4)×P(E4)] + D4 = 1.88;

E5 = C5 needs repair; O(E5) = 0.13; P(E5) = 0.12; D5 = 0.25; D5′ = [2×(1 − D5)×P(E5)] + D5 = 0.43;

O(H|E1&E2&E3&E4&E5) = O(H)×(A1′)×(A2′)×(D3′)×(D4′)×(D5′) = (1.50)×(1.30)×(1.42)×(1.09)×(1.88)×(0.43) = 2.44;

P(H|E1&E2&E3&E4&E5) = (2.44)/(3.44) = 0.71.

That is, P(sensible to repair the product) = 0.71.

If the cutoff value as decided by the decision maker is, say, 0.55, he will decide to send the used product for repair and for subsequent sale on a secondhand market.

11.5   FLEX-Based Expert System

An excellent tool called FLEX shell [1] is used to build an expert system that can decide if it is sensible to repair a particular used product for subsequent sale on the secondhand market. Figure 11.2 shows the user interface for executing the expert system.

In figure 11.2, the probability that it is sensible to repair the product is calculated by the FLEX-based expert system as 0.61. The difference in the probability obtained manually in section 11.4 and the one obtained by the expert system in this section is most likely due to the difference in the formulae used to calculate A′ and D′ (they are interpolated values). The user of an expert system shell cannot know how exactly the inference engine of the shell works. When there is a significant difference in the probability values, it is advisable to build the expert system using a knowledge representation language like Lisp or Prolog rather than an expert system shell.

11.6   Conclusions

In this chapter, using a numerical example, it is shown how an expert system can be built using Bayesian updating and fuzzy logic to decide whether it is sensible to repair an improperly functioning used product of interest for subsequent sale on a secondhand market instead of disassembling the product for subsequent reprocessing (remanufacturing or recycling).

```
WIN-PROLOG [Console]
File Edit Search Run Options Flex Window Help
| ?- second_hand(1, ... ).
Prob. : UPDATE    : (product is need_repair) = 1

Prob. : TRY       : sub_assly1
Prob. : LOOKUP    : (product is need_repair) = 1
Prob. : AFFIRMS   : weight(1.67) @ 1 -> 0.625468164794007
Prob. : UPDATE    : (s1 is need_repair) = 0.625468164794007
Prob. : FIRED     : sub_assly1

Prob. : TRY       : sub_assly2
Prob. : LOOKUP    : (product is need_repair) = 1
Prob. : AFFIRMS   : weight(2.5) @ 1 -> 0.714285714285714
Prob. : UPDATE    : (s2 is need_repair) = 0.714285714285714
Prob. : FIRED     : sub_assly2

Prob. : TRY       : comp1
Prob. : LOOKUP    : (s1 is need_repair) = 0.625468164794007
Prob. : AFFIRMS   : weight(2.22) @ 0.625468164794007 -> 0.566375418195992
Prob. : UPDATE    : (c1 is need_repair) = 0.566375418195992
Prob. : FIRED     : comp1

Prob. : TRY       : comp2
Prob. : LOOKUP    : (s1 is need_repair) = 0.625468164794007
Prob. : AFFIRMS   : weight(3.33) @ 0.625468164794007 -> 0.613105714886
Prob. : UPDATE    : (c2 is need_repair) = 0.613105714886
Prob. : FIRED     : comp2

Prob. : TRY       : comp3
Prob. : LOOKUP    : (s1 is need_repair) = 0.625468164794007
Prob. : AFFIRMS   : weight(1.82) @ 0.625468164794007 -> 0.546643121540394
Prob. : UPDATE    : (c3 is need_repair) = 0.546643121540394
Prob. : FIRED     : comp3

Prob. : TRY       : comp4
Prob. : LOOKUP    : (s2 is need_repair) = 0.714285714285714
Prob. : AFFIRMS   : weight(1.43) @ 0.714285714285714 -> 0.542184434270765
Prob. : UPDATE    : (c4 is need_repair) = 0.542184434270765
Prob. : FIRED     : comp4

Prob. : TRY       : comp5
Prob. : LOOKUP    : (s2 is need_repair) = 0.714285714285714
Prob. : AFFIRMS   : weight(1.25) @ 0.714285714285714 -> 0.525423728813559
Prob. : UPDATE    : (c5 is need_repair) = 0.525423728813559
Prob. : FIRED     : comp5

Prob. : TRY       : sens
Prob. : LOOKUP    : (c1 is need_repair) = 0.566375418195992
Prob. : AFFIRMS   : weight(3.5) @ 0.566375418195992 -> 0.571160931308024
Prob. : LOOKUP    : (c2 is need_repair) = 0.613105714886
Prob. : AFFIRMS   : weight(2) @ 0.613105714886 -> 0.550806141909616
Prob. : LOOKUP    : (c3 is need_repair) = 0.546643121540394
Prob. : AFFIRMS   : weight(0.75) @ 0.546643121540394 -> 0.494108820774045
Prob. : LOOKUP    : (c4 is need_repair) = 0.542184434270765
Prob. : AFFIRMS   : weight(0.13) @ 0.542184434270765 -> 0.480058661272842
Prob. : LOOKUP    : (c5 is need_repair) = 0.525423728813559
Prob. : AFFIRMS   : weight(2.12) @ 0.525423728813559 -> 0.513963258917539
Prob. : AND       : 0.480058661272842 * 0.513963258917539 -> 0.494908493010473
Prob. : AND       : 0.494108820774045 * 0.494908493010473 -> 0.489901864072248
Prob. : AND       : 0.550806141909616 * 0.489901864072248 -> 0.539905808194966
Prob. : AND       : 0.571160931308024 * 0.539905808194966 -> 0.609819395001277
Prob. : UPDATE    : (repair is sensible) = 0.609819395001277
Prob. : FIRED     : sens

Prob. : LOOKUP    : (s1 is need_repair) = 0.625468164794007
Prob. : LOOKUP    : (s2 is need_repair) = 0.714285714285714
Prob. : LOOKUP    : (c1 is need_repair) = 0.566375418195992
Prob. : LOOKUP    : (c2 is need_repair) = 0.613105714886
Prob. : LOOKUP    : (c3 is need_repair) = 0.546643121540394
Prob. : LOOKUP    : (c4 is need_repair) = 0.542184434270765
Prob. : LOOKUP    : (c5 is need_repair) = 0.525423728813559
Prob. : LOOKUP    : (repair is sensible) = 0.609819395001277
Sub1 = 0.625468164794007 ,
Sub2 = 0.714285714285714 ,
Com1 = 0.566375418195992 ,
Com2 = 0.613105714886 ,
Com3 = 0.546643121540394 ,
Com4 = 0.542184434270765 ,
Com5 = 0.525423728813559 ,
Sensib = 0.609819395001277
```

FIGURE 11.2
FLEX user interface for executing expert system.

References

1. Vasey, P., Westwood, D., and Johns, N. 1996. FLEX expert system toolkit. Logic Programming Associated Ltd.

12

Selection of New Products

12.1 The Issue

The focus of this issue is to help companies select and produce only those new products for which revenues in the closed-loop supply chain are expected to be higher than the costs. In this chapter, a cost-benefit function is formulated and then used to perform a multicriteria economic analysis for selecting an economical new product to produce in a closed-loop supply chain. The cost-benefit function can be defined as the ratio of the equivalent value of benefits associated with the object of interest to the equivalent value of costs associated with the same object. The equivalent value can be present worth, annual worth, future worth, etc. In this case, the object of interest is the new product to be produced in a closed-loop supply chain. The cost-benefit function (F) is formulated as

$$F = \frac{B}{C} \tag{12.1}$$

where B represents the equivalent value of the benefits (revenues) and C represents the equivalent value of the costs. An F value greater than 1.0 indicates that the object is economically advantageous. A notable point here is that due to uncertainties in supply, quality, and disassembly times in the reverse flow of the product (as a used product) in the closed-loop supply chain, decision makers must rely on experts' knowledge to obtain imprecise data for calculating B, C, and F values. Hence, fuzzy logic is used in the model presented in this chapter, and the cost-benefit function will hereafter be referred to as fuzzy cost-benefit function.

The fuzzy cost-benefit function consists of equivalent values of the following terms:

- New product sale revenue (revenue from selling new products, viz., products in the forward flow of the closed-loop supply chain)
- Reuse revenue (revenue from direct sale/usage in remanufacturing of usable components of used products)

- Recycle revenue (revenue from selling material obtained from recycling of unusable components of used products)
- New product production cost (cost to produce new products)
- Collection cost (cost to collect used products from consumers)
- Reprocessing cost (cost to remanufacture/recycle used products)
- Disposal cost (cost to dispose of the material left over after remanufacturing or recycling of used products)
- Loss-of-sale cost (cost due to loss of sale, which might occur occasionally due to lack of supply of used products)
- Investment cost (capital required for facilities and machinery involved in production of new products and collection and reprocessing of used products)

The chapter is organized as follows. In section 12.2, the assumptions made while formulating the fuzzy cost-benefit function are presented. In section 12.3, the nomenclature for the formulation of the fuzzy cost-benefit function is given. In section 12.4, the fuzzy cost-benefit function is formulated. In section 12.5, the model (economic analysis) for selecting an economical new product to produce in a closed-loop supply chain is presented. In section 12.6, a numerical example is given. Finally, section 12.7 provides some conclusions.

12.2   Assumptions

The following assumptions are made while formulating the fuzzy cost-benefit function:

1. The product of interest in the reverse flow (as a used product) of the closed-loop supply chain is completely disassembled.
2. All usable components of the product of interest in the reverse flow (as a used product) will be reused for direct sale or in remanufacturing, and all the remaining ones are recycled/disposed of.

12.3   Nomenclature

b_{ij} Probability of bad quality (broken, worn out, low performing, etc.) of component j in used product i

C_i Cost to produce one new product i

CC_i Total collection cost of used product i per period

CD Cost of reprocessing per unit time

CF	Recycling revenue factor ($/unit weight)
CR_i	Total recycle revenue of used product i per period
CO_i	Cost to collect one used product i
DC_i	Total disposal cost of used product i per period
D_i	Demand for new product i per period
DI_{ij}	Disposal cost index of component j in used product i (index scale 0 = lowest, 10 = highest)
DF	Disposal cost factor ($/unit weight)
E_{ik}	Subassembly k in product i
FCB_i	Fuzzy cost-benefit function for product i
i	Product type
IC_i	Investment cost of product i
j	Component type
LC_i	Loss-of-sale cost of used product i
M_i	Total number of subassemblies in product i
m_{ij}	Probability of missing component j in used product i
MC_i	Total production cost of new product i per period
N_{ij}	Multiplicity of component j in product i
RCP_{ij}	Percentage of recyclable contents by weight in component j of used product i
RC_i	Total reprocessing cost of used product i per period
RI_{ij}	Recycling revenue index of component j in used product i (index scale 0 = lowest, 10 = highest)
$Root_i$	Root node (for example, outer casing) of product i
RV_{ij}	Resale value of component j in used product i
SP_i	Selling price of new product i
SR_i	Total new product sale revenue of product i per period
SU_i	Supply of used product i per period
$T(Root_i)$	Time to disassemble $Root_i$
$T(E_{ik})$	Time to disassemble subassembly k in used product i
UR_i	Total reuse revenue of used product i per period
W_{ij}	Weight of component j in used product i
ΔBZ	Incremental total revenues (between the challenger and the defender)
ΔCZ	Incremental total costs (between the challenger and the defender)

12.4   Formulation of Fuzzy Cost-Benefit Function

The fuzzy cost-benefit function (*FCB*) of product i of interest consists of equivalent values (*EV*) of nine terms (total new product sale revenue per period (SR_i), total reuse revenue per period (UR_i), total recycle revenue per period (CR_i), total new product production cost per period (MC_i), total collection cost per period (CC_i), total reprocessing cost per period (RC_i), total disposal cost per period (DC_i), loss-of-sale cost (LC_i), and investment cost (IC_i)), as follows:

$$FCB_i = \frac{EV \text{ of } (SR_i + UR_i + CR_i)}{EV \text{ of } (MC_i + CC_i + RC_i + DC_i + LC_i + IC_i)} \qquad (12.2)$$

The following subsections explain how the above nine terms are calculated. Some of the terms are modified versions of those in [1].

12.4.1   Total New Product Sale Revenue per Period (*SR*)

SR of product i per period is influenced by the demand for new products per period (D_i) and the selling price of each new product (SP_i). This revenue equation can be written as follows:

$$SR_i = D_i.SP_i \qquad (12.3)$$

Often, in practice, objective data are available to express D_i and SP_i as crisp real numbers. Hence, SR_i is a crisp real number as well.

12.4.2   Total Reuse Revenue per Period (*UR*)

UR of product i is influenced by the fuzzy supply of the product per period (SU_i) and the following data of components of each type j in the product: the resale value (RV_{ij}), the number of components (N_{ij}), the fuzzy probability of missing (m_{ij}), and the fuzzy probability of bad quality (broken, worn out, low performing, etc.) (b_{ij}). This revenue equation can be written as follows:

$$UR_i = \sum_j SU_i.RV_{ij}.N_{ij}.(1 - b_{ij} - m_{ij}) \qquad (12.4)$$

Because SU_i, b_{ij}, and m_{ij} are expressed as fuzzy numbers, the resulting UR_i is a fuzzy number as well.

12.4.3   Total Recycle Revenue per Period (*CR*)

CR of product *i* is calculated by multiplying the component recycling revenue factors by the number of components recycled per period as follows:

$$CR_i = \sum_j \begin{bmatrix} SU_i.RI_{ij}.W_{ij}.RCP_{ij}. \\ \{N_{ij}(1-m_{ij}) - N_{ij}(1-b_{ij}-m_i)\} \end{bmatrix}.CF \qquad (12.5)$$

Note that each component has a percentage of recyclable contents (*RCP_{ij}*). RI_{ij} is the recycling revenue index (varying in value from 1 to 10) representing the degree of benefit generated by the recycling of component of type *j* (the higher the value of the index, the more profitable it is to recycle the component), W_{ij} is the weight of the component of type *j*, and *CF* is the recycling revenue factor. Because SU_i, b_{ij}, and m_{ij} are expressed as fuzzy numbers, the resulting CR_i is a fuzzy number as well.

12.4.4   Total New Product Production Cost per Period (*MC*)

MC of product *i* is calculated by multiplying the demand for new products per period (D_i) by the cost to produce one new product (C_i), as follows:

$$MC_i = D_i.C_i \qquad (12.6)$$

Often, in practice, objective data are available to express D_i and C_i as crisp real numbers. Hence, MC_i is a crisp real number as well.

12.4.5   Total Collection Cost per Period (CC)

CC of product *i* is calculated by multiplying the supply of used products per period (SU_i) by the cost of collecting one used product from consumers (CO_i), as follows:

$$CC_i = SU_i.CO_i \qquad (12.7)$$

Because SU_i is expressed as a fuzzy number, the resulting CC_i is a fuzzy number as well.

12.4.6   Total Reprocessing Cost per Period (*RC*)

RC of product *i* can be calculated from the supply of used products per period (SU_i), disassembly time of the root node (for example, outer casing) of the product ($T(Root_i)$), disassembly time of each subassembly in the product ($T(E_{ik})$), the reprocessing cost per unit time (*CD*), as follows:

$$RC_i = SU_i. \left[T(Root_i) + \sum_{k=1}^{M_i} T(E_{ik}) \right] .CD \qquad (12.8)$$

Depending upon the type (vague or objective) of data available for the disassembly times, RC_i is a fuzzy or crisp real number.

12.4.7 Total Disposal Cost per Period (*DC*)

DC of product *i* is calculated by multiplying the component disposal cost by the number of component units disposed per period, as follows:

$$DC_i = \sum_j \left[\begin{matrix} SU_i.DI_{ij}.W_{ij}.(1 - RCP_{ij}). \\ \{N_{ij}(1 - m_{ij}) - N_{ij}(1 - b_{ij} - m_{ij}) \end{matrix} \right] .DF \qquad (12.9)$$

Note that DI_{ij} is the disposal cost index (varying in value from 1 to 10) representing the degree of nuisance created by the disposal of components of type *j* (the higher the value of the index, the more nuisance the component creates, and hence it costs more to dispose it), W_{ij} is the weight of the component of type *j*, and *DF* is the disposal cost factor. Because SU_i, b_{ij}, and m_{ij} are expressed as fuzzy numbers, the resulting CR_i is a fuzzy number as well.

12.4.8 Loss-of-Sale Cost per Period (*LC*)

LC of product *i* represents the cost of not meeting its demand in a timely manner. This occurs because of the unpredictable supply of used products, as consumers do not discard them in a predictable manner. *LC* is difficult to predict and thus is usually guessed by experts. Due to the involvement of the experts' guesses, LC_i is expressed as a fuzzy number.

12.4.9 Investment Cost (*IC*)

IC of product *i* is the capital required for facilities and machinery involved in production of new products and collection and reprocessing of used products. Depending upon the type (vague or objective) of data available for the product and the location of the facilities, IC_i is a fuzzy or crisp real number.

12.5 Model

In order to select the most economical new product to produce in a closed-loop supply chain, from a set of candidate products, the following steps are used:

Step 1: Eliminate every candidate product whose *FCB* is less than 1.0.

Step 2: Assign the candidate product that has the lowest *IC* as the defender and the product with the next-lowest *IC* as the challenger.

Step 3: Calculate the ratio of the EV of incremental total revenue ΔBZ (between the challenger and the defender) to the EV of incremental total cost ΔCZ (between the challenger and the defender). If the ratio is less than 1.0, eliminate the challenger. Otherwise, eliminate the defender.

Step 4: Repeat steps 2 and 3 until only one product (which is the most economical one in the set) is left.

12.6 Numerical Example

Three different products (products 1–3) whose structures are shown in figures 12.1–12.3, respectively, are considered for the example.

It is assumed that the supplies of all these products are perpetual. Hence, capitalized worth (CW) [2] is taken as the EV. Therefore, *FCB* is the ratio of CW of total revenues to CW of total costs. The data necessary to calculate *FCB* of products 1–3 are given in tables 12.1–12.2 and 12.3, respectively.

Also, $T(Root_1) = 2$ min; $T(Root_2) = 1.5$ min; $T(Root_3) = 1.5$ min; $T(E_{11}) = 9$ min; $T(E_{21}) = 7$ min; $T(E_{22}) = 8$ min; $T(E_{31}) = 7$ min; $T(E_{32}) = 8$ min; $SU_1 = (200, 230, 250)$ products per year; $SU_2 = (210, 220, 230)$ products per year; $SU_3 = (600, 650, 700)$ products per year; $CO_1 = \$20$; $CO_2 = \$21$; $CO_3 = \$18$; $IC_1 = \$20,000$; $IC_2 = \$25,000$; $IC_3 = \$30,000$; $D_1 = 900$ products per year; $D_2 = 850$ products per year; $D_3 = 1,000$ products per year; $SP_1 = \$70$; $SP_2 = \$28$; $SP_3 = \$58$; $C_1 = \$25$; $C_2 = \$30$; $C_3 = \$28$; $LC_1 = \$(300, 500, 700)$ per year; $LC_2 = \$(100, 400, 500)$ per year; $LC_3 = \$(900, 1,000, 1,100)$ per year; $CF = 0.2$ \$/lb; $DF = 0.1$ \$/lb; and $CD = 0.55$ \$/min.

Upon calculating revenues and costs for each product, one gets $FCB_1 = (2.13, 2.45, 2.88)$, $FCB_2 = (0.77, 0.84, 0.91)$, and $FCB_3 = (1.76, 2.11, 2.68)$. Defuzzifying these numbers using equation (4.8), one gets $FCB_1 = 2.48$, $FCB_2 = 0.94$, and $FCB_3 = 2.17$. Because FCB_2 is less than 1.0, product 2 is eliminated from further analysis.

Now, because IC_1 is less than IC_3, product 1 is considered the defender and product 3 is considered the challenger. The defuzzified ratio of CW of ΔBZ to CW of ΔCZ is now calculated as 1.86, which is greater than 1.0. Hence, the defender (product 1) is eliminated. Therefore, the remaining product (product 3) is the most economical new product (among the three products) to produce in the closed-loop supply chain.

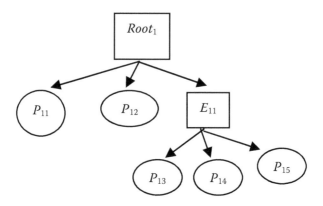

FIGURE 12.1
Structure of product 1.

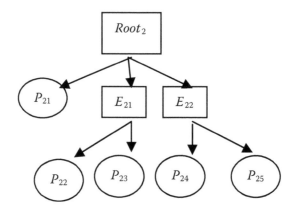

FIGURE 12.2
Structure of product 2.

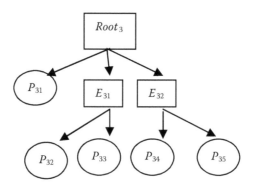

FIGURE 12.3
Structure of product 3.

TABLE 12.1

Data of product 1

Component	RV_{1j} ($)	N_{1j}	W_{1j} (lb)	RI_{1j}	RCP_{1j}	DI_{1j}	b_{1j}	m_{1j}
P_{11}	7.0	3	4.5	5	65%	6	(0.1,0.1, 0.2)	(0.3, 0.4, 0.4)
P_{12}	8.0	4	6.5	5	50%	4	(0.5, 0.6, 0.7)	(0.1, 0.2, 0.2)
P_{13}	9.0	2	7.0	3	75%	4	(0.2, 0.3, 0.4)	(0.3, 0.4, 0.4)
P_{14}	6.9	1	2.7	9	35%	5	(0.2, 0.2, 0.3)	(0.1, 0.1, 0.2)
P_{15}	8.4	5	7.5	6	70%	1	(0.1, 0.1, 0.2)	(0.3, 0.4, 0.5)

TABLE 12.2

Data of product 2

Component	RV_{2j} ($)	N_{2j}	W_{2j} (lb)	RI_{2j}	RCP_{2j}	DI_{2j}	b_{2j}	m_{2j}
P_{21}	1.0	1	3.9	2	40%	3	(0.1, 0.1, 0.2)	(0.0, 0.1, 0.1)
P_{22}	1.5	3	1.5	4	20%	1	(0.1, 0.2, 0.2)	(0.0, 0.0, 0.0)
P_{23}	1.2	7	4.1	1	70%	2	(0.2, 0.3, 0.4)	(0.2, 0.2, 0.3)
P_{24}	2.5	4	3.2	5	90%	4	(0.3, 0.4, 0.5)	(0.1, 0.1, 0.2)
P_{25}	3.1	3	2.0	2	50%	2	(0.3, 0.4, 0.4)	(0.1, 0.1, 0.2)

TABLE 12.3

Data of product 3

Component	RV_{3j} ($)	N_{3j}	W_{3j} (lb)	RI_{3j}	RCP_{3j}	DI_{3j}	b_{3j}	m_{3j}
P_{31}	9.0	2	4.0	9	30%	4	(0.2, 0.3, 0.4)	(0.1, 0.2, 0.3)
P_{32}	8.0	5	5.0	7	60%	3	(0.1, 0.2, 0.2)	(0.1, 0.2, 0.2)
P_{33}	9.0	3	2.0	8	70%	1	(0.3, 0.4, 0.4)	(0.1, 0.2, 0.2)
P_{34}	7.0	2	6.0	9	25%	3	(0.2, 0.3, 0.3)	(0.3, 0.3, 0.4)
P_{35}	7.0	1	5.2	6	50%	2	(0.3, 0.3, 0.4)	(0.1, 0.1, 0.2)

12.7 Conclusions

In this chapter, a cost-benefit function is formulated and then used to perform a multicriteria economic analysis for selecting an economical new product to produce in a closed-loop supply chain. Fuzzy logic is used in formulating the cost-benefit function.

References

1. Veerakamolmal, P., and Gupta, S. M. 1999. Analysis of design efficiency for the disassembly of modular electronic products. *Journal of Electronics Manufacturing* 9:79–95.
2. Sullivan, W. G., Wicks, E. M., and Luxhoj, J. T. 2003. *Engineering economy*. 12th ed. Englewood Cliffs, NJ: Prentice Hall.

13

Selection of Secondhand Markets

13.1 The Issue

In chapter 11, an expert system is built using Bayesian updating and fuzzy logic to decide whether it is sensible to repair a used product of interest for subsequent sale on a secondhand market. In this chapter, fuzzy logic, quality function deployment (QFD), and method of total preferences are used to select the market with the most potential to sell a used product in, from a set of candidate secondhand markets.

This chapter is organized as follows: Section 13.2 gives the performance aspects and enablers for the application of QFD. Section 13.3, using a numerical example, presents the implementation of the model that uses fuzzy logic, quality function deployment (QFD), and method of total preferences to select the market with the most potential to sell a used product in, from a set of candidate secondhand markets. Finally, section 13.4 gives some conclusions.

13.2 Performance Aspects and Enablers for Application of QFD

The following are the performance aspects of the secondhand markets,* which are considered in the application of QFD:

1. Before-sale performance (BSP) (reflects the ability to attract new customers to the secondhand market)
2. While-sale performance (WSP) (reflects the ability to motivate the customers to buy secondhand products while the customers are in the secondhand market)

---

* It should be noted that a secondhand market here means a "store" where secondhand (i.e., used) products are sold, along with new products.

3. After-sale performance (*ASP*) (reflects the ability to attract old customers to the secondhand market)

The following are the enablers considered in the application of QFD:

- Good advertisement (*AD*)
- High difference of prices between new and secondhand products (*DP*)
- Greenness of the sale (*GS*)
- Incentives (warranty, service, etc.) (*IC*)
- Low average price of products (*LP*)
- Good location of sale of secondhand products (placement in front of the new ones, etc.) (*LS*)
- Proper maintenance of secondhand products (as proper as it is for new products) (*MN*)
- Discounts to returning customers (*RC*)
- Good return/exchange policy (*RP*)
- Reputation of the store (*RS*)
- Variety of secondhand products on the shelves (*VS*)

The enablers for *BSP* are *LP, IC, RS,* and *AD*; those for *WSP* are *DP, GS, IC, MN, AD, VS, RP,* and *LS*; and those for *ASP* are *RC* and *AD*.

13.3   Selection of Potential Secondhand Markets

A numerical example is used to present the approach as follows. Two secondhand markets are compared, and the one that has more potential than the other is to be selected. Suppose that there are five experts (M's) for giving linguistic values to R_{ij} and d_i data. Table 13.1 shows the linguistic scale for R_{ij} as well as for d_i.

Tables 13.2–13.4 show the linguistic relationship scores (R_{ij}) as given by the five experts to the enablers of *BSP, WSP,* and *ASP*, respectively.

Table 13.5 gives the linguistic importance values (d_i) of *BSP, WSP,* and *ASP,* as given by the five experts.

The ATIRs and RTIRs of the enablers are then calculated using equations (4.30) and (4.31), respectively. It must be noted that the triangular fuzzy numbers (TFNs) are averaged before defuzzifying. For example, consider *AD*, which is an enabler for three performance aspects, *BSP, WSP,* and *ASP*. The linguistic relationship scores, as given by the five experts, for *AD* and *BSP* (see table 13.2), for *AD* and *WSP* (see table 13.3), and for *AD* and *ASP* (see table 13.4) are converted into TFNs using table 13.1. These TFNs are then

TABLE 13.1
Conversion table for linguistic R_{ij} data

Linguistic R_{ij}	Triangular fuzzy number (TFN)
Very strong (VS)	(7.5. 10, 10)
Strong (S)	(5, 7.5, 10)
Medium (M)	(2.5, 5, 7.5)
Weak (W)	(0, 2.5, 5)
Very weak (VW)	(0, 0, 2.5)

TABLE 13.2
Linguistic relationship scores of *BSP* and its enablers

	M1	M2	M3	M4	M5
LP	VS	VS	M	W	S
IC	S	S	M	VS	S
RS	M	S	VS	W	M
AD	VS	VS	VS	VS	VS

TABLE 13.3
Linguistic relationship scores of *WSP* and its enablers

	M1	M2	M3	M4	M5
DP	V	VS	M	W	S
GS	N	M	VS	VS	S
IC	S	S	M	VS	S
MN	M	S	VS	W	M
AD	S	S	VS	M	W
VS	VS	VS	VS	VS	VS
RP	S	VS	M	VS	S
LS	M	S	VS	W	N

TABLE 13.4
Linguistic relationship scores of *ASP* and its enablers

	M1	M2	M3	M4	M5
RC	VS	VS	M	W	S
AD	VW	S	M	S	VS

averaged as follows: The average relationship score for *AD* and *BSP*, as calculated using equations (4.4) and (4.7), is

$$\left(\frac{7.5+7.5+7.5+7.5+7.5}{5}, \frac{10+10+10+10+10}{5}, \frac{10+10+10+10+10}{5}\right)$$
$$= (7.5, 10, 10)$$

Similarly, the average relationship scores for *AD* and *WSP* and *AD* and *ASP* are calculated as (4, 6, 8.5) and (4, 6, 8), respectively. Defuzzified average relationship scores for *AD* and *BSP*, *AD* and *WSP*, and *AD* and *ASP*, as calculated using equation (4.8), are 9.17, 6.33, and 6, respectively. The linguistic importance values (d_j) of *BSP*, *WSP*, and *ASP*, as given by the five experts (see table 13.5), are converted into TFNs using table 13.1 again. The average importance values for *BSP*, *WSP*, and *ASP* are then calculated as (6.5, 9, 10) (defuzzified value = 8.5), (5, 7.5, 9.5) (defuzzified value = 7.33), and (5.5, 8, 9.5) (defuzzified value = 7.67), respectively. ATIR of *AD* is then calculated using equation (4.30) as (8.5)×(9.17) + (7.33)×(6.33) + (7.67)×(6) = 212.79. The ATIRs of all the enablers are shown in table 13.6. RTIR of each enabler is then calcu-

TABLE 13.5
Linguistic importance values of performance aspects

	M1	M2	M3	M4	M5
BSP	VS	VS	VS	S	S
WSP	S	S	M	S	VS
ASP	VS	M	S	S	VS

TABLE 13.6
ATIRs and RTIRs of enablers

Enabler	ATIR	RTIR
AD	170.36	0.24
DP	48.89	0.07
GS	46.44	0.06
IC	116.11	0.16
LP	56.67	0.08
LS	36.67	0.05
MN	36.67	0.05
RC	51.11	0.07
RP	42.50	0.06
RS	42.50	0.06
VS	67.22	0.10
Sum	715.14	1.00

lated using equation (4.31). For example, RTIR of AD is the ratio of its ATIR

to the sum of the ATIRs of all the enablers, i.e., $\dfrac{170.36}{715.14}$ = 0.24. RTIRs are also shown in table 13.6.

Table 13.7 shows the scale for converting the linguistic WA_{nj} value given by the five experts (for arithmetic simplicity, we assume here that there is a consensus among the experts). The WA_{nj} (linguistic) values and the corresponding defuzzified TFNs (calculated using equation (4.8)) for each secondhand market are shown in table 13.8.

Then, using equation (4.32), the TUPs for the two secondhand markets are calculated. For example, the TUP for secondhand market 1 is calculated using the RTIR$_j$ values (see table 13.7) and the defuzzified WA_{1j} values (see table 13.8) as follows: (0.24)×(7.33) + (0.07)×(1) + (0.06)×(3) + (0.16)×(7.33) + (0.08)×(7.33) + (0.05)×(9) + (0.05)×(5) + (0.07)×(1) + (0.06)×(7.33) + (0.06)×(1) + (0.09)×(3) = 5.35. Similarly, the TUP for secondhand market 2 is calculated as 6.74.

TABLE 13.7

Conversion table for linguistic WA_{nj} data

Linguistic WA$_{nj}$	TFN
Very good (VG)	(7, 10, 10)
Good (G)	(5, 7, 10)
Fair (F)	(2, 5, 8)
Poor (P)	(1, 3, 5)
Very poor (VP)	(0, 0, 3)

TABLE 13.8

Linguistic and corresponding defuzzified *WA* values of secondhand markets

Enabler	Market 1 Linguistic WA$_{nj}$	TFN	Market 2 Linguistic WA$_{nj}$	TFN
AD	G	7.33	G	7.33
DP	VP	1	VG	9
GS	P	3	VG	9
IC	G	7.33	VG	9
LP	G	7.33	G	7.33
LS	VG	9	VP	1
MN	F	5	P	3
RC	VP	1	F	5
RP	G	7.33	G	7.33
RS	VP	1	VP	1
VS	P	3	G	7.33

Finally, using equation (4.33), NTUPs are calculated for the two secondhand markets. For example, NTUP for secondhand market 1 is calculated as follows:

$$\frac{5.35}{5.35 + 6.74} = 0.44$$

Similarly, NTUP for secondhand market 2 is calculated as 0.56.

It is evident that secondhand market 2 has more potential than secondhand market 1.

13.4   Conclusions

In this chapter, fuzzy logic, quality function deployment (QFD), and method of total preferences are used to select the market with the most potential to sell a used product in, from a set of candidate secondhand markets.

14

Design of a Synchronized Reverse Supply Chain

14.1 The Issue

Effective management of a supply chain requires synchronization among the internal business processes of the supply chain. Synchronization in a supply chain means reducing the variability among the internal business processes or partners such that each stakeholder in the supply chain acts in a way that is appropriately timed with the actions of the other stakeholders [1]. The delivery performance of a supply chain is maximized largely by synchronizing the internal business processes (reducing variability) such that the final product fits in the customer-specified delivery window with a very high probability.

In this chapter, a model consisting of two design experiments that use the Six Sigma concept to achieve better synchronization in a reverse supply chain is presented. This model tailors the individual processes in such a way that the overall delivery performance is maximized.

The chapter is organized as follows: section 14.2 presents the model, and section 14.3 gives some conclusions.

14.2 Model (Two Design Experiments)

In this section, two design experiments are presented. The first experiment (section 14.2.1) determines the range of nominal values (for a Six Sigma delivery performance) for the lead times of different individual processes of a reverse supply chain. The second experiment (section 14.2.2) determines a variance pool for the lead times of the different individual processes.

14.2.1 First Experiment (Determination of Nominal Pool)

Six processes of a reverse supply chain are considered:

1. *Procurement*: Procurement involves obtaining the used products from the consumers at the collection centers.
2. *Inspection/testing*: Inspection/testing involves determining the condition of the products collected in order to determine whether to remanufacture, refurbish, or recycle the product.
3. *Disassembly*
4. *Remanufacture/refurbish*
5. *Transportation*: Involves transporting the remanufactured products to the markets.
6. *Delivery*

It is assumed that the supply chain deals only with remanufacturing or refurbishing of the used products (a third-party recycler takes care of the products meant for recycling). It is also assumed that there is no waiting time between the six processes. Hence, the nominal value of the reverse supply chain's lead time, τ_y, is described as the sum of the nominal values of lead times of the individual processes:

$$\tau_y = \sum_{i=1}^{6} \tau_i \tag{14.1}$$

Also [2],

$$\sigma_Y^2 = \sum_{i=1}^{n} \sigma_i^2 ; \sigma_i = \frac{T_i}{3C_{pki}} \tag{14.2}$$

where σ_i and T_i are the standard deviations and tolerances of the individual processes and σ_Y is the overall process standard deviation.

In finding a nominal pool, the tolerances for the lead times of the individual processes of the reverse supply chain are given in this experiment, as are the nominal values of the lead times of some of the processes (in this experiment, τ_2, τ_3, τ_4). The nominal value of the lead time of the reverse supply chain, as well as its tolerance, is also known. The problem (see table 14.1) is now to find a range of values for the other nominal values (in this case, τ_1, τ_5, τ_6), so as to achieve a Six Sigma delivery performance. Note that this has its implications on the choice of suppliers, carriers, and other logistics providers.

Let $\tau_y = 100$ days, $T_y = 12$ days, $T_1 = 3$ days, $T_2 = 4$ days, $T_3 = 1$ day, $T_4 = 2$ days, $T_5 = 2$ days, and $T_6 = 1$ day. Also, let the nominal values of inspection/testing, disassembly, and remanufacturing/refurbishing be 20, 25, and 30 days, respectively. It is now required to find a range of values for the pool of other nominal values ($\tau_1 + \tau_5 + \tau_6$) such that the probability of delivery is at least 0.9999966 within the delivery window, which is ($\tau_y - T_y, \tau_y + T_y$) = (88, 112).

Suppose that the individual processes are Six Sigma processes, which implies a C_{pk} value of 1.5 for each process. Using equation (14.2), the standard

TABLE 14.1

Finding a nominal pool (first experiment)

Given:

τ_y = nominal value (target) of the lead time of the reverse supply chain, y
T_y = tolerance range of the lead time of the reverse supply chain, y
T_i (i = 1, 2, ..., 6) = tolerances of the lead times of the individual processes
τ_2, τ_3, τ_4 = nominal values of lead times of individual processes 2, 3, and 4

To compute:
A range of values for each of the nominal values, τ_1, τ_5, τ_6, over which Six Sigma delivery performance is guaranteed

deviations (σ) of the individual processes as well as the reverse supply chain are found to be $\sigma_1 = 0.67$, $\sigma_2 = 0.87$, $\sigma_3 = 0.22$, $\sigma_4 = 0.44$, $\sigma_5 = 0.44$, $\sigma_6 = 0.22$, and $\sigma_y = 1.299$. For example,

$$\sigma_1 = \frac{T_1}{3C_{pk1}} = \frac{3}{3 \times 1.5} = 0.67$$

Also, from equation (14.1), one can see that $(\tau_1 + \tau_5 + \tau_6) = \tau_y - (\tau_2 + \tau_3 + \tau_4) = 100 - 75 = 25$. From equation (14.2), the overall standard deviation for the lead times of the three processes (1, 5, and 6) is $\sigma_{156} = \sqrt{\sigma_1^2 + \sigma_5^2 + \sigma_6^2} = 0.99945$. For the set of these three processes to have Six Sigma performance, a shift of at most 1.5 standard deviations from the target (= 25) is acceptable. Hence, as long as the nominal value of the lead time of the set falls in the range (τ_{156} ± $1.5\sigma_{156}$), Six Sigma delivery performance is guaranteed. In other words, if $(\tau_1 + \tau_5 + \tau_6)$ is in the range (23.5, 26.5), Six Sigma delivery performance is guaranteed. That is, for this range of values, the probability of y to be in the range (88, 112) is at least 0.9999966. The maximum probability is attained at 25 days.

14.2.2 Second Experiment (Determination of Variance Pool)

In this experiment, the nominal values of the lead times of the individual processes considered in the reverse supply chain are given, as are the tolerances for the lead times of some of the processes. The nominal value of the lead time of the reverse supply chain, as well as its tolerance is also known. The problem now is to find a variance pool that can be distributed across the individual processes whose tolerances are not known (see table 14.2).

Let $\tau_y = 100$ days, $T_y = 12$ days; $\tau_1 = 20$ days, $\tau_2 = 5$ days, $\tau_3 = 25$ days, $\tau_4 = 30$ days, $\tau_5 = 12$ days, and $\tau_6 = 8$ days. Also, let the tolerance of the lead time of procurement, disassembly, and remanufacturing be 3 days, 2 days, and 1 day, respectively.

Here, too, it is assumed that the individual processes are Six Sigma processes, which implies a C_{pk} value of 1.5 for each process. Also, a Six Sigma

TABLE 14.2

Finding a variance pool (second experiment)

Given:

τ_y = nominal value (target) of the lead time of the reverse supply chain, y

T_y = tolerance range of the lead time of the reverse supply chain, y

τ_i (i = 1, 2, ..., 6) = nominal values of the lead times of the individual processes

T_1, T_3, T_4 = tolerances of lead times of the individual processes 1, 3, and 4

To compute:

A variance pool that can be distributed across the individual processes whose tolerances are
  not known

delivery performance implies C_p for the reverse supply chain = 2 and C_{pk} for the reverse supply chain = 1.5. Using equation (14.2), the standard deviations of the individual processes whose tolerances are known (procurement, disassembly, and remanufacturing, in this experiment) as well as the reverse supply chain are found to be $\sigma_1 = 2$, $\sigma_3 = 0.44$, $\sigma_4 = 0.22$, and $\sigma_y = 2.67$. Again, using equation (14.2), one can get the variance pool $(\sigma_2^2 + \sigma_5^2 + \sigma_6^2)$, for the three processes (inspection/testing, transportation, and delivery) whose tolerances are unknown (inspection/testing, transportation, and delivery in this example) as 2.89. This variance pool can now be distributed among the individual processes based on engineering judgment that ensures a Six Sigma delivery performance for the reverse supply chain. That is, for this variance pool, the probability of y to be in the range (88, 112) is at least 0.9999966.

14.3   Conclusions

The delivery performance of a supply chain is maximized by synchronizing the internal business processes such that the final product fits in the customer-specified delivery window with a very high probability. In this chapter, a model consisting of two design experiments that use the Six Sigma concept to achieve better synchronization in a reverse supply chain was presented. Six individual processes—procurement, inspection/testing, disassembly, remanufacture/refurbish, transportation, and delivery—were considered in the supply chain. The model tailors the individual processes in such a way that the overall delivery performance of the reverse supply chain is maximized.

References

1. Antony, J., Swarnkar, R., and Tiwari, M. K. 2006. Design of synchronized supply chains: A genetic algorithm based Six Sigma constrained approach. *International Journal of Logistics Systems and Management* 2:120–41.
2. Narahari, Y., Viswanadham, N., and Bhattacharya, R. 2000. Design of synchronized supply chains: A Six Sigma tolerancing approach. *IEEE International Conference on Robotics and Automation* 1151–56.

15

Performance Measurement

15.1 The Issue

In the era of globalization of markets and business process outsourcing, many firms realize the importance of continuous monitoring of their supply chain's performance for its effectiveness and efficiency [1]. Traditionally, performance measurement is defined as the process of quantifying the effectiveness and efficiency of action [2]. In the modern era, performance measurement has a far more significant role than just quantification and accounting. It provides management with important feedback to monitor performance, reveal progress, diagnose problems, and enhance transparency among the several tiers of the supply chain, thus making a phenomenal contribution to decision making, particularly in redesigning business goals and reengineering processes [3, 4].

Developing the performance measurement systems is a difficult aspect of performance measure selection. It involves the methods by which an organization creates its measurement system. Important questions that need to be addressed include [5]: What to measure? How often to measure? How are multiple individual measures integrated into a measurement system? Also, the way each performance aspect is weighed is industry specific [6]. For example, although customer satisfaction is an indication of the standard level of service, for different industries, customers look at different measures. By the same token, due to the inherent differences in various aspects between the forward and reverse supply chains (see chapter 1), the performance aspects and evaluation techniques used in a forward supply chain cannot be extended to a reverse/closed-loop supply chain.

In this chapter, appropriate performance aspects and their enablers (drivers of performance aspects) are identified for a reverse/closed-loop supply chain environment, and a performance measurement model that uses linear physical programming (LPP) and quality function deployment (QFD) is presented.

The chapter is organized as follows. Section 15.2 describes the application of LPP to QFD optimization. Section 15.3 enlists the performance aspects and enablers identified for a reverse/closed-loop supply chain environment and then illustrates the model using a numerical example. Finally, section 15.4 gives some conclusions.

15.2   Application of LPP to QFD Optimization

Applying LPP to QFD optimization involves two steps [6]. The first step involves collecting information for LPP through the traditional QFD approach and building the house of quality (HOQ); this step helps the design team to complete a qualitative analysis of the design problem. The second step involves mathematical modeling using LPP. The complete process is illustrated in figure 15.1.

15.2.1   First Step

The house of quality (HOQ) shows the customer requirements, engineering characteristics, and the competitor's performance analysis (see figure 15.2). Some engineering characteristics are interrelated; as a result, changing

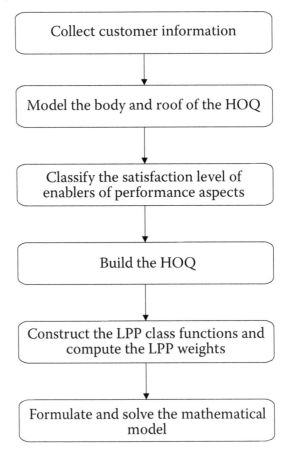

FIGURE 15.1
Application of LPP to QFD optimization.

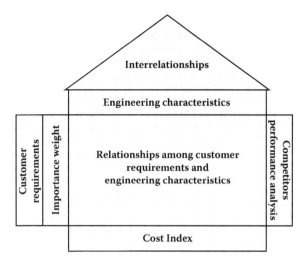

FIGURE 15.2
House of Quality

the value of one engineering characteristic may alter its impact on customer requirements or on other engineering characteristics. Notice that the roof of the HOQ shows the interrelationships among the engineering characteristics. These relationships are based on the knowledge and experience of the design team. In this chapter, the notation γ_{jk}, which denotes the elements of the correlation matrix, is introduced to describe the correlation between the jth and kth engineering characteristics. The body of the HOQ shows the relationships between engineering characteristics and customer requirements. R_{ik} denotes the relationship between the ith customer requirement and jth engineering characteristic. These relationships are expressed on a scale of 1–9, where 1 denotes a weak relation and 9 denotes the strongest relationship. In order to quantify the impact of the dependencies between the engineering characteristics on the relationship between the customer requirements, the model transformation shown in equation (15.1) is used:

$$R_{ij}^{norm} = \frac{\sum_{k=1}^{n} R_{ik}\gamma_{kj}}{\sum_{j=1}^{n}\sum_{k=1}^{n} R_{ij}\gamma_{jk}}, i=1,2,...,m; j=1,2,...,n$$

(15.1)

where R_{ij}^{norm} can be interpreted as the incremental change in the level of fulfillment of the ith customer requirement when the jth engineering characteristic is satisfied to a certain level; m is the number of customer requirements and n is the number of engineering characteristics.

Using LPP, the satisfaction level of each customer requirement is classified in one of the six different ranges: ideal, desirable, tolerable, undesirable, highly undesirable, and unacceptable. Each objective is described by one of the eight subclasses, four soft and four hard.

The roof of the HOQ contains the interrelationships between the engineering characteristics, denoted by γ_{jk}, and the body represents the interrelationships between the customer requirements and the engineering characteristics, denoted by R_{ij}^{norm}.

The bottom of the HOQ shows the cost index for each engineering characteristic. In most cases, improvements in engineering characteristics will result in an increase in cost. X_j ($j = 1, 2, ..., n$) is defined as the value of engineering characteristic j, and the maximum value of the engineering characteristic is max $\{X_j\}$. Then the normalized value of engineering characteristic j is defined as

$$x_j = X_j/\max\{X_j\}, j = 1, 2, ..., n \tag{15.2}$$

Suppose that when c_j is invested in engineering characteristic j, it can achieve its maximum value, max $\{X_j\}$; however, because of constraints, in most cases not all engineering characteristics can achieve their maximum value. To make engineering characteristic j achieve x_j, the design team needs to invest $c_j \times x_j$ in engineering characteristic j, thus making the cost function a monotonically increasing one.

15.2.2   Second Step

The aim of the design team would be to attain the highest customer satisfaction level while meeting the budget limitations. The LPP problem is formulated as follows:

$$Minimize \sum_{i=1}^{m} \sum_{s=2}^{5} \left[\tilde{w}_{is}\, d_{ps}^{-} + \tilde{w}_{is}\, d_{ps}^{+} \right] \tag{15.3}$$

where d_{ps} represents the deviations of the customer requirements from their target values, subject to:

$$\sum_{j=1}^{n} R_{ij}^{norm} - d_{is}^{+} \leq t_{i(s-1)}^{+} \quad \text{and} \quad \sum_{j=1}^{n} R_{ij}^{norm} \leq t_{i5}^{+} \tag{15.4}$$

for all i in classes $1S, 3S, 4S, i = 1, 2, .., n_{sc}; s = 2, .., 5;$

$$\sum_{j=1}^{n} R_{ij}^{norm} + d_{is}^{-} \geq t_{i(s-1)}^{-} \quad and \quad \sum_{j=1}^{n} R_{ij}^{norm} \geq t_{i5}^{-}$$

(15.5)

for all i in classes $2S, 3S, 4S, i = 1, 2, .., n_{sc}; s = 2, .., 5;$

$$B \geq \sum_{j=1}^{n} c_j x_j$$

(15.6)

where B is the maximum available budget;

$$d_{is}^{+} \geq 0; d_{is}^{-} \geq 0$$

(15.7)

$$0 \leq x_j \leq 1$$

(15.8)

Note that in the above formulation, total budget is the only system constraint; however, depending on the decision environment, other constraints, such as minimum satisfaction level for the ith customer requirement or maximum achievement level for the jth engineering characteristic, can be added.

15.3 Reverse/Closed-Loop Supply Chain Performance Measurement

In this section, performance aspects and their enablers for measuring the performance of a reverse/closed-loop supply chain are identified, and then the model that uses LPP and QFD for performance measurement is presented. Note that in the LPP–QFD methodology described in section 15.2, the engineering characteristics are replaced by performance aspects and the customer requirements are replaced by enablers.

15.3.1 Performance Aspects and Enablers

The following are the performance aspects considered in the model:

1. *Reputation*: This aspect deals with the firm's overall reputation in the industry. It is driven by enablers such as on-time delivery ratio, returning customers ratio, and the firm's green image. Green image

can be measured in a number of different ways, such as the environmental expenses incurred, the amount of waste disposed of, or the fairs/symposiums related to environmentally conscious manufacturing in which the firm participates. In this model, the amount of waste disposed of is used to measure the green image.

2. *Innovation and improvement*: This aspect is driven by enablers such as the R&D expenses ratio or the number of new products and processes launched.

3. *Public participation*: This aspect is a measure of the firm's marketing capabilities that may be driven by enablers such as flexibility, or the firm's ability to handle uncertainties, after-sales service efficiency, and the markets targeted (markets targeted is quantified as the percentage of customers who are knowledgeable about the firm's product; the more knowledgeable the customers are, the better the participation will be). After-sales service efficiency can be defined as the ratio of the number of customers served to the number of customers seeking service.

4. *Facility potentiality*: Potentiality of the remanufacturing facilities can be driven by several enablers, such as the location of the facilities, the increment in the quality of products (quality of outgoing reprocessed products–quality of incoming used products), disassembly time multiplied by throughput, throughput divided by the supply of used products (see section 7.2.1 for a detailed explanation of these enablers), and usage of automated disassembly systems. Location can be quantified by specifying it as the distance from the nearest major city, and usage of automated disassembly systems can be quantified using the investment on automated disassembly systems.

5. *Responsiveness*: This aspect reflects how well the firm responds to the ever-changing customer specifications and can be driven by enablers such as flexibility, the firm's ability to handle uncertainties, and the firm's after-sales service efficiency. Flexibility can be quantified using the number of different varieties of products the firm manufactures/remanufactures.

6. *Delivery reliability*: This aspect reflects how well the firm meets the due dates specified by the customers and can be driven by enablers such as the effectiveness of the firm's master production schedule, the usage of automated disassembly systems, the supply of used products, and the quality of used products. The effectiveness of the firm's master production schedule can be defined as the ratio of the number of orders delivered no later than the due date to the total number of orders delivered.

Other operational aspects, such as resource utilization, cost per operating hour, and manufacturing lead time, are also considered in the model.

15.3.2 Numerical Example

In this section, LPP–QFD methodology detailed in section 15.2 is illustrated using a numerical example. Table 15.1 shows the LPP preference ranges for the enablers that constitute the right-hand side of the HOQ.

Table 15.2 shows the relationships between the performance aspects, denoted as γ_{jk}, evaluated on a 1 to 9 scale.

Table 15.3 shows the relationships between the ith enabler and the jth performance aspect, denoted as R_{ij}.

Using equation (15.1), the relationships between performance aspects and enablers are normalized and the normalized scores form the body of the HOQ. Table 15.4 shows the HOQ for the performance measurement. For simplicity, the roof has been removed after normalizing R_{ij}.

TABLE 15.1

LPP preference ranges for enablers

Enabler	LPP class	Unit	Ideal	Desirable	Tolerable	Highly undesirable	Unacceptable
On-time delivery ratio	2S	%	100	80	75	65	50
Returning customers ratio	2S	%	100	90	80	75	65
Green image	1S	$	0	1,200	1,500	2,000	2,500
R&D expenses ratio	4S	%	50	45	40	35	30
			60	65	75	80	100
New products/ processes	4S	Unit	20	15	10	8	5
			25	30	35	40	45
Flexibility	2S	Unit	65	55	50	40	35
After-sales service efficiency	2S	%	100	85	75	60	50
Markets targeted	2S	%	100	80	75	60	55
Throughput × Disassembly time (TP × DT)	2S	%	100	90	80	75	65
Throughput/ supply of used products (TP/SU)	2S	%	100	95	85	70	65
Quality of reprocessed products–quality of used products (QO–QI)	2S	%	65	60	50	45	40
Effectiveness of MPS	2S	%	100	85	75	70	65
Facility location	4S	Unit	35	30	25	20	15
			20	25	35	40	50
Labor cost per hour	1S	$	20	25	35	40	50
Usage of automated DA systems	4S	$	1,000	750	600	500	250
			2,000	2,500	3,500	4,000	5,000

TABLE 15.2

Relationships between performance aspects (γ_{jk})

Performance aspect	Reputation	I&I	PP	RES	FP	DR	RU	C/Oh	MLT
Reputation	—	7	8	7	6	8	4	3	7
Innovation & improvement (*I&I*)	7	—	6	4	4	3	3	3	2
Public participation (*PP*)	8	6	—	7	3	2	2	3	2
Responsiveness (*RES*)	7	4	7	—	3	4	3	3	2
Facility potentiality (*FP*)	6	4	3	3	—	6	8	8	8
Delivery reliability (*DR*)	8	3	2	4	6	—	7	5	5
Resource utilization (*RU*)	4	3	2	3	8	7	—	6	8
Cost per operating hour (*C/Oh*)	3	3	3	2	8	5	6	—	5
Manufacturing lead time (*MLT*)	7	2	2	2	8	5	8	5	—

TABLE 15.3

Relationships between performance aspects and enablers (R_{ij})

	Reputation	I&I	PP	RES	FP	DR	RU	C/Oh	MLT
On-time delivery ratio	7	4	5	4	6	9	6	5	7
Returning customers ratio	7	5	5	7	5	8	6	4	6
Green image	8	3	4	7	6	4	6	3	3
R&D expenses ratio	4	7	3	5	4	5	5	6	3
New products/processes	6	7	4	4	5	4	4	3	4
Flexibility	7	4	7	8	6	5	4	3	4
After-sales service efficiency	8	4	8	8	4	4	3	5	3
Markets targeted	5	3	7	5	3	5	3	5	3
TP×DT	5	3	5	4	8	8	8	7	8
TP/SU	5	3	5	4	8	8	8	7	8
QO–QI	7	3	6	5	9	7	8	7	7
Effectiveness of MPS	4	2	4	5	5	8	7	5	7
Facility location	5	3	6	5	4	6	5	7	5
Labor cost per hour	5	2	5	4	6	6	6	8	5
Usage of automated DA systems	5	7	5	4	8	8	8	8	8

TABLE 15.4

House of quality for performance measurement

Performance aspects	Reputation	I&I	PP	RES	FP	DR	RU	C/Oh	MLT	t_1	t_2	t_3	t_4	t_5
Enablers														
On-time delivery ratio	0.138	0.091	0.087	0.093	0.134	0.114	0.12	0.106	0.114	100	80	75	65	50
Returning customers ratio	0.142	0.092	0.097	0.091	0.131	0.101	0.116	0.881	0.111	100	90	80	75	65
Green image	0.117	0.093	0.096	0.084	0.112	0.102	0.098	0.098	0.113	0	1,200	1,500	2,000	2,500
R&D expenses ratio	0.128	0.075	0.091	0.078	0.12	0.101	0.102	0.093	0.102	50	45	40	35	30
										60	65	75	80	100
New products/processes	0.122	0.078	0.09	0.085	0.109	0.095	0.102	0.082	0.098	20	15	10	8	5
										25	30	35	40	45
Flexibility	0.128	0.091	0.092	0.085	0.105	0.106	0.1	0.083	0.098	65	55	50	40	35
After-sales service efficiency	0.122	0.083	0.08	0.088	0.109	0.106	0.096	0.104	0.095	100	85	75	60	50
Markets targeted	0.128	0.083	0.075	0.0884	0.115	0.109	0.1	0.102	0.096	100	80	75	60	55
TP×DT	0.13	0.087	0.078	0.082	0.126	0.104	0.116	0.1	0.181	100	90	80	75	65
TP/SU	0.135	0.084	0.08	0.08	0.126	0.109	0.116	0.102	0.11	100	95	85	70	65
QO–QI	0.124	0.087	0.078	0.079	0.12	0.104	0.111	0.1	0.111	65	60	50	45	40
Effectiveness of MPS	0.135	0.088	0.075	0.083	0.131	0.102	0.115	0.102	0.107	100	85	75	70	65
Facility location	0.125	0.076	0.079	0.078	0.126	0.102	0.108	0.088	0.102	35	30	25	20	15
										20	25	35	40	50
Labor cost	0.125	0.09	0.082	0.079	0.121	0.108	0.113	0.094	0.112	20	25	35	40	50
Usage of automated DA systems	0.13	0.094	0.078	0.081	0.124	0.103	0.112	0.097	0.105	1,000	750	600	500	250
Cost index	4,000	2,500	2,000	2,500	2,500	3,000	2,000	2,000	2,000					

TABLE 15.5

Normalized LPP weights of enablers

Enablers	Weights			
On-time delivery ratio	$\tilde{w}_{12}^{-} = 0.000068$	$\tilde{w}_{13}^{-} = 0.00414$	$\tilde{w}_{14}^{-} = 0.028$	$\tilde{w}_{15}^{-} = 0.967$
Returning customers ratio	$\tilde{w}_{22}^{-} = 0.000224$	$\tilde{w}_{23}^{-} = 0.0034$	$\tilde{w}_{24}^{-} = 0.034$	$\tilde{w}_{25}^{-} = 0.956$
Green image	$\tilde{w}_{32}^{+} = 0.000114$	$\tilde{w}_{33}^{+} = 0.006$	$\tilde{w}_{34}^{+} = 0.05$	$\tilde{w}_{35}^{+} = 0.93$
R&D expenses ratio	$\tilde{w}_{42}^{-} = 0.0002$	$\tilde{w}_{43}^{-} = 0.003$	$\tilde{w}_{44}^{-} = 0.06$	$\tilde{w}_{45}^{-} = 0.93$
	$\tilde{w}_{42}^{+} = 0.001$	$\tilde{w}_{43}^{+} = 0.0073$	$\tilde{w}_{44}^{+} = 0.251$	$\tilde{w}_{45}^{+} = 0.74$
New products/processes	$\tilde{w}_{52}^{-} = 0.00016$	$\tilde{w}_{53}^{-} = 0.0023$	$\tilde{w}_{54}^{-} = 0.094$	$\tilde{w}_{55}^{-} = 0.9$
	$\tilde{w}_{52}^{+} = 0.00027$	$\tilde{w}_{53}^{+} = 0.0034$	$\tilde{w}_{54}^{+} = 0.06$	$\tilde{w}_{55}^{+} = 0.935$
Flexibility	$\tilde{w}_{62}^{-} = 0.00041$	$\tilde{w}_{63}^{-} = 0.0122$	$\tilde{w}_{64}^{-} = 0.084$	$\tilde{w}_{65}^{-} = 0.902$
After-sales service efficiency	$\tilde{w}_{72}^{-} = 0.000183$	$\tilde{w}_{73}^{-} = 0.0043$	$\tilde{w}_{74}^{-} = 0.039$	$\tilde{w}_{75}^{-} = 0.956$
Markets targeted	$\tilde{w}_{82}^{-} = 0.00006$	$\tilde{w}_{83}^{-} = 0.00414$	$\tilde{w}_{84}^{-} = 0.0174$	$\tilde{w}_{85}^{-} = 0.97$
TP×DT	$\tilde{w}_{92}^{-} = 0.000274$	$\tilde{w}_{93}^{-} = 0.0039$	$\tilde{w}_{94}^{-} = 0.125$	$\tilde{w}_{95}^{-} = 0.87$
TP/SU	$\tilde{w}_{102}^{-} = 0.00028$	$\tilde{w}_{103}^{-} = 0.0018$	$\tilde{w}_{104}^{-} = 0.019$	$\tilde{w}_{105}^{-} = 0.97$
QO–QI	$\tilde{w}_{112}^{-} = 0.00028$	$\tilde{w}_{113}^{-} = 0.0018$	$\tilde{w}_{114}^{-} = 0.06$	$\tilde{w}_{115}^{-} = 0.935$
Effectiveness of MPS	$\tilde{w}_{122}^{-} = 0.00009$	$\tilde{w}_{123}^{-} = 0.002$	$\tilde{w}_{124}^{-} = 0.06$	$\tilde{w}_{125}^{-} = 0.935$
Facility location	$\tilde{w}_{132}^{-} = 0.000274$	$\tilde{w}_{133}^{-} = 0.0039$	$\tilde{w}_{134}^{-} = 0.06$	$\tilde{w}_{135}^{-} = 0.93$
	$\tilde{w}_{132}^{+} = 0.00054$	$\tilde{w}_{133}^{+} = 0.0036$	$\tilde{w}_{134}^{+} = 0.125$	$\tilde{w}_{135}^{+} = 0.87$

(continued)

TABLE 15.5 (continued)

Normalized LPP weights of enablers

Enablers	Weights			
Labor cost per hour	$\tilde{w}_{142}^{+} = 0.0006$	$\tilde{w}_{143}^{+} = 0.006$	$\tilde{w}_{144}^{+} = 0.05$	$\tilde{w}_{145}^{+} = 0.94$
Usage of automated DA systems	$\tilde{w}_{152}^{-} = 0.0002$	$\tilde{w}_{153}^{-} = 0.006$	$\tilde{w}_{154}^{-} = 0.155$	$\tilde{w}_{155}^{-} = 0.83$
	$\tilde{w}_{152}^{+} = 0.0005$	$\tilde{w}_{153}^{+} = 0.003$	$\tilde{w}_{154}^{+} = 0.125$	$\tilde{w}_{155}^{+} = 0.87$

Table 15.5 shows the LPP weights of the enablers.

The model is formulated as detailed in section 15.2.3. Apart from the total budget constraint, we consider three other constraints (see equations (15.9)–(15.11)). Equations (15.9) and (15.10) limit the maximum achievement levels of performance aspects, facility potentiality, and resource utilization to realistic values of 0.85 and 0.8, respectively. Equation (15.11) sets the minimum level for the enabler, on-time delivery ratio, to 0.65. It is assumed that the total budget in this example is 19,000 monetary units.

$$x_{potentiality} \leq 0.85 \tag{15.9}$$

$$x_{ResourceUtilization} \leq 0.8 \tag{15.10}$$

$$\sum_{j} R^{norm}_{Ontime\ Delivery\ Ratio,j} x_j \geq 0.65 \tag{15.11}$$

The model is solved using LINGO (v8), and tables 15.6 and 15.7 show the respective results. Table 15.6 shows the budget allocation and the achievement levels of performance aspects, and table 15.7 shows the satisfaction levels of the enablers

$$(= \sum_{j} R^{norm}_{ij} x_j).$$

From table 15.6, it is evident that the satisfaction levels of the enablers are almost the same. This can be attributed to the fact that they are coordinated with the preference of LPP that always puts more effort into the aspects that lag behind.

TABLE 15.6

Budget allocation and achievement levels of performance aspects

	Reputation	I&I	Public participation	Responsiveness	Facility potentiality	Delivery reliability	Resource utilization	Cost/Oh	Manufacturing lead time
Budget allocation	1,272	2,500	2,000	2,500	2,125	2,700	1,600	2,000	1,520
Achievement level (x_j)	0.318	1.0	1.0	1.0	0.85	0.9	0.8	1.0	0.76

TABLE 15.7

Enabler satisfaction levels

Enabler	Satisfaction level (%)
On-time delivery ratio	82.21
Returning customers ratio	82.18
Green image	77.27
R&D expenses ratio	72.84
New products/processes	73.35
Flexibility	73.74
After-sales service efficiency	73.07
Markets targeted	72.41
$TP{\times}DT$	81.82
TP/SU	76.42
$QO{-}QI$	76.53
Effectiveness of MPS	76.73
Facility location	74.49
Labor cost per hour	76.20
Usage of automated DA systems	74.28

15.4 Conclusions

In this chapter, aspects and enablers (drivers of aspects) for measurement of performance of a reverse/closed-loop supply chain were identified, and then a model that uses linear physical programming and quality function deployment for performance measurement was presented.

References

1. Gunasekaran, A., Patel, C. and Tirtiroglu, E. 2001. Performance measures and metrics in a supply chain environment. *International Journal of Operations and Production Management* 21:71–87.
2. Neely, A., Gregory, M., and Platts, K. 1995. Performance measurement system design: A literature review and research agenda. *International Journal of Operations and Production Management* 15:80–116.
3. Rolstandas, A. 1995. *Performance measurement: A business process benchmarking approach*. New York: Chapman & Hall.
4. Waggoner, D. B., Neely, A. D., and Kennerley, M. P. 1999. The forces that shape organizational performance measurement systems: An interdisciplinary review. *International Journal of Production Economics* 60:53–60.
5. Beamon, M. B. 1999. Measuring supply chain performance. *International Journal of Operations and Production Management* 19:275–92.
6. Chan, F. T. S. 2003. Performance measurement in a supply chain. *International Journal of Advanced Manufacturing Technology* 21:534–48.

16

Conclusions

The growing desire of consumers to acquire the latest technology, both at home and in the workplace, along with the rapid technological development of new products, has led to a new environmental problem: waste, viz., used products that are discarded prematurely. But as the saying goes that behind every problem lies an opportunity, reprocessing of this waste means

1. *Saving natural resources*: We conserve land and reduce the need to drill for oil and dig for minerals by making products using materials and components obtained from reprocessing instead of virgin materials.
2. *Saving energy*: It usually takes less energy to make products from reprocessed materials and components than from virgin materials.
3. *Saving clean air and water*: Making products from reprocessed materials and components creates less air pollution and water pollution than from virgin materials.
4. *Saving landfill space*: When reprocessed materials and components are used to make a product, they do not go into landfills.
5. *Saving money*: It costs much less to make products from reprocessed materials and components than from virgin materials.

Besides the above opportunities, an important driver for companies to engage in reprocessing is the enforcement of environmental regulations by local governments.

A reverse supply chain consists of a series of activities required to collect used products from consumers and reprocess them (used products) to either recover their leftover market values or properly dispose of them. Today, in practice, it has become common for companies involved in a forward supply chain (series of activities required to produce new products from virgin materials and distribute the former to consumers) to also carry out collection and reprocessing of used products. This combined practice of forward and reverse supply chains is called a closed-loop supply chain. In the past decade, there has been an explosive growth of reverse and closed-loop supply chains in both scope and scale.

Strategic planning (also called designing) primarily involves the structuring (which products should be processed/produced in which facilities) of a supply chain over the next several years. It is long-range planning and is typically performed every few years when a supply chain needs to expand its capabilities. The issues faced by strategic planners of reverse and closed-loop supply chains are:

- Selection of used products
- Evaluation of collection centers
- Evaluation of recovery facilities
- Optimization of transportation of goods
- Evaluation of marketing strategies
- Evaluation of production facilities
- Evaluation of futurity of used products
- Selection of new products
- Selection of secondhand markets
- Synchronization of supply chain processes
- Supply chain performance measurement

In all of these issues, strategic planners must meet the following challenges: uncertainty in supply rate of used products, unknown condition of used products, and imperfect correlation between supply of used products and demand for reprocessed goods.

This book addressed the above issues amidst the above challenges in a variety of decision-making situations using efficient models. These models implement several quantitative techniques, such as:

- Analytic hierarchy process
- Eigen vector method
- Analytic network process
- Fuzzy logic
- Extent analysis method
- Fuzzy multicriteria analysis method
- Quality function deployment
- Method of total preferences
- Linear physical programming
- Goal programming
- Technique for order preference by similarity to ideal solution (TOPSIS)
- Borda's choice rule
- Expert systems
- Bayesian updating
- Taguchi loss function
- Six Sigma

- Neural networks
- Geographical information systems
- Linear integer programming

The issues addressed in this book can serve as foundations for other researchers to build bodies of knowledge in this new and fast-growing field of research, i.e., strategic planning of reverse and closed-loop supply chains. Furthermore, the models proposed in this book for those issues can be utilized by industrialists for understanding how a particular issue in the strategic planning of reverse and closed-loop supply chains can be effectively approached in a particular decision-making situation, using a suitable quantitative technique or a suitable combination of two or more quantitative techniques.

Author index

Subject Index

A

absolute technical importance ratings (ATIR) 48–49, 190, 192, 246, 248–249

actor 29

affirms weight 59, 60

after-sale performance (ASP) 16, 246–248

alternatives 37–39, 44–48, 51, 55–58, 80, 126

analytic hierarchy process (AHP) 6, 8, 37–39, 43–45, 50, 125–126, 151, 272

analytic network process (ANP) 6, 8, 25, 37, 39, 43–44, 88, 110–111, 116–118, 124, 182, 192, 198, 201, 214, 226, 272

aspiration level 53–54, 117, 174

assembly 19–21, 127

attitude 45–46, 202, 204, 222

attitude of management (AMT) 202, 204

average disassembly time of used products 127

average quality of used products 127

average supply of used products 127

B

Bayesian updating 7–8, 37, 59, 227–232, 245, 272

before-sale performance (BSP) 16, 245–248

Borda score 58, 101–102, 109, 140, 147

Borda's choice rule 7, 8, 37, 58, 88, 95, 101, 103, 109, 124–125, 135, 140–141, 147, 151, 272

branch and bound (B&B) 29

by-products 204

C

capacity planning 27

carpet 25–26

case study 28, 30

class functions 50–51, 258

class 1H 52, 158

class 1S 50–51, 78, 131, 158, 168

class 2H 52

class 2S 78–79, 131, 169

class 3H 52

class 3S 50

class 4H 52

class 4S 50

classical numerical constraints 12, 73

cleaning 25

closed-loop supply chain 3–9, 11–17, 19–21, 24–25, 27, 29, 31, 37, 87, 161, 167–168, 173, 179, 181, 188, 192, 198, 201–202, 211, 226, 235–236, 241, 243, 257, 261, 269, 271, 273

collection center(s) 3–6, 8, 11–14, 25, 87–95, 97–111, 115–125, 128, 131–132, 154–158, 161–162, 164–171, 173, 175–178, 182, 188–189, 201–202, 204–205, 252, 272

collection cost 15, 74–76, 78, 236, 238–239

comparative importance values 38, 40, 112–113, 119

competition 5

complete disassembly 15, 19, 227, 236

connection weight(s) 68, 87, 104, 125, 143

consistency ratio (CR) 38, 112, 119, 128

constraint(s) 8, 12–14, 21–22, 26–27, 52, 55, 69, 73, 117, 123–124, 153, 157, 159–160, 163–164, 166, 168–169, 171, 174, 176, 178–179, 260, 261, 267

construction waste 25

consumer(s) 1, 3, 4, 12, 14–15, 30–31, 73, 76, 78, 87–89, 93–109, 134–147, 157, 167, 181, 236, 239–240, 252, 271

convenience 93, 134, 182

cooperation 111, 119, 182, 189, 194

copier 28

cost-minimization 26

cost-benefit function 8, 24, 74–75, 235–238, 243

crisp performance matrix 46, 222–224

customer service 23, 88, 128, 130, 132, 202, 204–205

For Product Safety Concerns and Information please contact our EU
representative GPSR@taylorandfrancis.com
Taylor & Francis Verlag GmbH, Kaufingerstraße 24, 80331 München, Germany

www.ingramcontent.com/pod-product-compliance
Ingram Content Group UK Ltd.
Pitfield, Milton Keynes, MK11 3LW, UK
UKHW021619240425
457818UK00018B/651